METALLOMICS
A Primer of Integrated Biometal Sciences

METALLOMICS
A Primer of Integrated Biometal Sciences

Wolfgang Maret

Kings's College London, UK

ICP Imperial College Press

Published by

Imperial College Press
57 Shelton Street
Covent Garden
London WC2H 9HE

Distributed by

World Scientific Publishing Co. Pte. Ltd.
5 Toh Tuck Link, Singapore 596224
USA office: 27 Warren Street, Suite 401-402, Hackensack, NJ 07601
UK office: 57 Shelton Street, Covent Garden, London WC2H 9HE

Library of Congress Cataloging-in-Publication Data
Names: Maret, W. (Wolfgang), author.
Title: Metallomics : a primer of integrated biometal sciences / Wolfgang Maret.
Description: [London] ; New Jersey : Imperial College Press, [2016] |
    Includes bibliographical references and index.
Identifiers: LCCN 2015048800| ISBN 9781783268276 (hc : alk. paper) |
    ISBN 9781783268283 (sc : alk. paper)
Subjects: | MESH: Metals--metabolism | Trace Elements | Metalloproteins--physiology |
    Metabolomics
Classification: LCC QP532 | NLM QU 130.2 | DDC 612.3/924--dc23
LC record available at http://lccn.loc.gov/2015048800

British Library Cataloguing-in-Publication Data
A catalogue record for this book is available from the British Library.

Desk Editors: Harini/Mary Simpson

Typeset by Stallion Press
Email: enquiries@stallionpress.com

Contents

Prologue

The book was written with the intent of drawing wider attention to the importance of *metal ions in the biological sciences.* It introduces *Metallomics* as an integrated biometals science and discusses how a significant number of essential and non-essential chemical elements are controlled and interact in functional units such as a biological cell. It is not an introduction to methodologies — it focuses on what has been achieved, not on how it has been achieved. Because Metallomics is an emerging field, the book necessarily presents a snapshot of our emerging knowledge.

More than just a few elements determine the biochemistry of life. In fact, at least 20 are needed in humans, 11 of which are metals. This fact is not widely appreciated because of a general misconception that substances of lower abundance (trace elements) are thought to be qualitatively inferior to those of higher abundance (bulk elements). The book embraces a wider view of the biochemistry of metal ions as it also discusses the interaction of non-essential metals with life. It emphasizes what biochemistry, the chemistry of life processes, gains when it is based on the entire periodic system of the elements instead of just a few chemical elements. It tries to overcome the perception that metallobiochemistry is a specialized field. Biochemistry includes organic as well as inorganic chemistry.

Many books on bioinorganic chemistry have been written (Lippard and Berg 1994; Frausto da Silva and Williams 2001; Bertini *et al.* 2007; Crichton 2012; Kaim *et al.* 2013; Rehder 2014). They deal with the structures and functions of metal ion complexes with biological ligands from a chemical perspective. Yet other books focus on the biology or medicine of metals, among them a classical text with the first four editions written by the late Eric J. Underwood (Mertz 1986). The present book is different in several regards from all these books and it is not a substitute for these books to which the interested reader is referred. It bridges chemistry and biology and provides a different entry point to the field. It does not simply provide a parts list or functions of individual metal ions in biology. Rather it discusses how individual components interact and relate to overall functions in biological systems. *The biological perspective teaches us that structures of metal complexes gain functional significance in the biological context. Often, it encompasses functions that are not readily apparent from the chemistry of the isolated complexes.* The theme pervading the book is that metals should not be treated simply as cofactors in biochemistry but as critical, integral parts of cellular processes. The book reaches out to other disciplines by discussing the significance of metal ions for nutrition, toxicology, pharmacology, medicine, and ecology — the interaction of biological systems with their environments.

Metallomics is not just another name or "the emperor's new clothes". It generates new opportunities for understanding the roles of metal ions in biology. Many metal ions are present simultaneously in a living system and this fact limits which structures and functions they can adopt. Metallomics is an integrative science which tries to overcome the division of investigations of metal ions into many scientific communities, some of which focus on just one metal ion, e.g. iron, copper, zinc. As a consequence of this separation different terminologies have evolved. One would hope that the field of Metallomics will bring communities together and thus make it clear to the wider scientific community how important metals (and non-metals) are and that metallobiochemistry is a fundamental and significant part of biochemistry. Accepting this premise would seem to require a different starting point for teaching biochemistry.

The book introduces beginners to the field with some fundamental concepts but also prepares for advanced studies because it leads the reader to the questions at the frontiers in a developing field. It describes work in progress. It aspires to stimulate inquiries, and indicates opportunities for careers in an area of research where there is potential for significant discoveries in the future. Metallomics has the advantage of developing breadth in addition to depth of understanding due to its interdisciplinary and transdisciplinary nature.

I thank my colleagues from the Metal Metabolism Group and the London Iron Metabolism Group at King's College London, in particular Christer Hogstrand and Robert Hider for providing a stimulating research atmosphere, my teachers Michael Zeppezauer (Saarland University, Germany), Marvin W. Makinen (The University of Chicago), and the late Bert L. Vallee (Harvard Medical School), Fred Madsen for discussing metals in nutrition, but also others who are remembered and whose encounters had long-lasting effects on my scientific development: Sir Hans Krebs, Helmut Beinert, F. Albert Cotton, Robert "Bob" J. P. Williams, and Ivano Bertini, and many other mentors, friends and colleagues whose infective enthusiasm about progress in the natural sciences in general and metal ions in biology in particular had a significant influence on my attempts to stay on course and on strengthening my belief in the bright future of the field of biometals.

References

D. Rehder, Bioinorganic Chemistry. Oxford University Press 2014.

I. Bertini, H.B. Gray, E.I. Stiefel, and J. Silverstone Valentine, Biological Inorganic Chemistry: Structure and Reactivity. University Science Books 2007.

J.J.R. Frausto da Silva and R.J.P. Williams, The Biological Chemistry of the Elements. The Inorganic Chemistry of Life. Oxford University Press 2001.

R. Crichton, Biological Inorganic Chemistry, 2nd edition. A new introduction to molecular structure and function. Elsevier, Amsterdam 2012.

S.J. Lippard and J.M. Berg, Principles of Bioinorganic Chemistry. University Science Books 1994.

W. Kaim, B. Schwederski, and A. Klein, Bioinorganic Chemistry: Inorganic Elements in the Chemistry of Life. Wiley 2013.

W. Mertz, ed., Trace Elements in Human and Animal Nutrition, 5th edition, volumes 1&2, Academic Press 1986.

List of abbreviations

ATP: adenosine triphosphate
ER: endoplasmic reticulum
GTP: guanosine triphosphate
HMW: high molecular weight
LMW: low molecular weight
MT: metallothionein
MTF-1:   metal regulatory element (MRE)-binding transcription factor-1
PSE: periodic system of the elements
SOD: superoxide dismutase

Chapter 1

Introduction

Contemporary biosciences are based on the molecules of life built from the main elements: carbon, hydrogen, oxygen, nitrogen, sulfur, and phosphorus. By and large, the roles of other elements essential for life or interacting with life are much less appreciated or sometimes even neglected altogether. The book tries to close this gap with a focus on the usage and significance of metals in biology. It explains why the functions of a large number of chemical elements should be a fundamental aspect of the chemistry of life (biochemistry) rather than a specialty discipline.

Joens Jacob Berzelius (1779–1848), one of the founders of modern chemistry, wrote in his textbook of chemistry (1827):

> "In living nature the elements appear to obey quite different laws from those in dead nature. Hence quite different products result from their various interactions than in inorganic nature".

His writing indicated, presciently, that there is something special in the way the elements are utilized in life. With considerable vision, he added the word "appear" because, after all, the principles of chemistry and physics apply in living nature. Friedrich Wöhler (1800–1882), who spent some time in the laboratory of Berzelius, synthesized urea in 1828 and with this feat is credited of having started the chemistry of life, organic chemistry, and of having begun to end the debate surrounding vitalism, namely the belief that a vital force is necessary for the things to live.

The origins of our understanding of the functions of the elements date back to ground-breaking work in the 19th century when some metal ions were found to be nutritionally essential for life. The field developed significantly over the course of the last century with additional metals being discovered to be essential. However, how many chemical elements are essential for life continues to be a matter that is not settled. Initially, trace element research was largely dominated by developing methodologies to measure metal ions in biological material, which was not straightforward due to the limited number of methods available and their poor sensitivity. Functional context was often established much later by associating a deficiency or an overload of a metal ion with physiological or pathophysiological events, and finally molecular actions. New technologies have lowered the detection limits for metal ions significantly and allowed data collection with ever increasing resolution in biological time and space. These advances are driving new discoveries. Renewed interest of chemists in the second half of the last century established the discipline known as bioinorganic chemistry and inorganic biochemistry, or metallobiochemistry if the focus is on metals as in this book. The late Helmut Beinert (1913–2007), a pioneer in the field, wrote about *bioinorganic chemistry*: "Definitions of fields are becoming blurred, and we must recognize the fallacy of trying to categorize with any rigidity while still preserving real meaning." (Beinert 2002). The blurring is inherent in the interpretation of the names for the disciplines. Bioinorganic chemistry may include areas of purely chemical investigations of biomolecules, which may not be relevant to how metals function in biology. An interaction of a protein with a metal *in vitro* is not evidence of the interaction occurring *in vivo* and does not reflect the entire context of biological function that is becoming apparent only in the biological system and its environment. Metallobiochemistry loses its biological context if it fails to address the question: Does this chemistry occur in biology? Furthermore, identifying biochemistry with organic chemistry resulted in the unfortunate misunderstanding, propagated in most textbooks and curricula, that inorganic biochemistry is a specialty discipline. It is not. Organic chemistry is generally associated with the chemistry of the main elements of life, carbon, hydrogen, oxygen, and nitrogen, whereas inorganic chemistry is seen as the chemistry of the rest of the elements in non-living matter. But life is not possible without the participation of a significant number of additional elements. Distinguishing organic from inorganic chemistry in biology does not foster an understanding of biochemistry as the chemistry of life. This book tries to restore this balance by basing biochemistry on all the elements essential for life and those interacting with life.

The book embraces a biological perspective, examining the structures and functions of metal ions in living systems and the biological context. It leads to insights that are not apparent from the individual molecules in isolation. Metallobiochemistry certainly is a field in its own right and a subdiscipline of biochemistry at the same level as others that address the building blocks of life: sugars, lipids, amino acids, and nucleotides. Furthermore, the treatment of metal ions in biochemistry textbooks has been largely confined to their molecular functions as cofactors without reference to their wider role in biological control of cells and significance in the physiology of organisms. Metals are more than just spurious "trace elements". Often they are discussed side-by-side with vitamins, which are also essential and serve as cofactors or hormones. But metals are different. They are elements of life with more fundamental and pleiotropic functions than those of the water-soluble vitamins.

Like biochemistry, metallobiochemistry intersects with many other disciplines. Additional words such as biological inorganic chemistry, inorganic cell biology, inorganic chemical biology, and inorganic physiology have been introduced to restore or expand the relationship of bioinorganic chemistry to biology, but what is needed is a full integration into biochemistry and molecular and cellular biology. Why do we then need another word to describe the discipline dealing with metals in biology, especially one where a relationship to biochemistry is not in the name and thus perhaps blurred again? The answer is that *metallomics* adds to the scope of investigations in the way it approaches the subject differently and offers a higher level of integration. A major difference between metallomics and bioinorganic chemistry is that the latter focuses on interactions of metal ions with isolated biomolecules whereas the former describes the roles of metal ions in a biological system, where all other metal ions and biomolecules are also present. Bioinorganic chemistry is part of metallomics, but the approaches are different. Similar to metallobiochemistry, metallomics is comprised of several approaches and knowledge from different disciplines. However, it focuses on systems rather than individual parts. This focus has already extended the horizon of the field of biometals considerably: (i) it provided estimates of the number of metalloproteins and increased their number by at least an order of magnitude, (ii) it demonstrated that many more metals than the ones currently discussed to be associated with biological structure are present and thus relevant for biological function, and (iii) it identified and characterized novel functions of proteins in the control of metal metabolism as a result of asking how metals function in the biological systems. The latter point requires further explanations. Once one addresses metal ions in a biological system, one realizes that the simultaneous presence of metal ions in the system sets constraints on the availability and usage of different metals ions and that intricate control is required to ascertain specificity and to avoid unwanted interactions. This holistic approach made us aware of new protein functions and mechanisms in the process of how organisms control the availability of metal ions. Clearly the whole is more than the sum of its parts.

The remarks do not belittle the enormous contributions of bioinorganic chemistry, which has been and remains a very exciting endeavour in revealing fascinating metal coordination environments in proteins and uncovering the role of metal ions in chemically challenging reactions. Once structures of biological metal coordination environments became known, they often had no precedent in known chemistry. This sequence of events does not surprise if one considers that the efforts of synthetic chemists cover only about 200 years while Nature had millions of years to find solutions to performing challenging chemical reactions at ambient temperature and pressure.

Unique coordination environments in proteins occur not only in catalytic, structural and regulatory metal centers but also in other metalloproteins that function in handling metal ions with specific mechanisms for binding, transport, and storage (Maret and Wedd 2014). Thus, biology has found chemical solutions to acquire metals, controlling them, and utilizing them to control metabolism and signalling, importantly all *with selectivity for individual metal ions in a situation where many other metal ions are present simultaneously* (Williams and Fraústo da Silva 2000). This functionally and structurally rich metallobiochemistry of living systems provides significant feedback for theoretical and synthetic concepts in chemistry, in part realized in the discipline biomimetic chemistry, where newly discovered principles of biological inorganic chemistry are exploited in synthetic chemistry.

The word "metallomics" stands for an approach and methods as well as for a discipline. The book addresses the discipline, the roles of metal ions in life. It does not address all forms of life, however. It focuses on bacteria, from which a lot of knowledge on metalloenzymes and mechanisms of metal control was garnered, and on humans, because we want to understand the importance of metals for our health and diseases. The reader should keep in mind the incredible variety of bacteria, plants and animals and hence variations in usage of biometals. Another reason for a focus on the two living systems, bacteria and humans, is that an understanding of the symbiotic or parasitic relationship between the two is also important for our health and disease.

References

H. Beinert (2002). Bioinorganic chemistry: A new field? *J. Biol. Chem.* 277, 37967–37972.

R.J.P. Williams and J.J.R. Fraústo da Silva (2000). The distribution of elements in cells. *Coord. Chem. Rev.* 200–202, 247–348.

W. Maret and A. Wedd, eds., Binding, Transport, and Storage of Metal Ions in Biological Cells, Royal Society of Chemistry Press, Cambridge, United Kingdom 2014.

Chapter 2

Metallomics

2.1 Metals, the metallome, and metallomics

Scientists apply conceptually different approaches. They are either "analytical" (top-down), establishing a parts list, or "synthetic" (bottom-up), assembling parts to functional units. Metallomics embraces both approaches. It employs methods for the determination and speciation of metal ions (a reductionist approach) and addresses the question how metal ions work in a system (a holistic approach). This book focuses on metals in biological systems and *the cell as the main biological unit of organisms.*

Metallomics is not just another fad among fashionable 'omics' pursuits. It has a place of its own side-by-side with other important 'omes' that refer to the characterization of the molecular building blocks of life: genomes, proteomes, lipidomes, and glycomes (Figure 2.1).

Strictly, metallomics focuses on metals. The term "elementomics" has been employed to expand the discipline to non-metal elements. However, not the elemental metals but the metal ions are important in biology though a distinction in this regard is rarely made in the bioscience literature. To address this distinction, the term "ionomics" has been used. The term itself does not draw attention to metals and for the purpose of the investigation of biometals is useful in combination with naming the element, e.g. calcium ionomics.

The late Robert JP Williams (1926–2015) coined the term "metallome", describing all the metals and their species in a system (Williams 2001). Hiroki Haraguchi is credited for coining the word "metallomics". While travelling with highly inspired scientists on a high-spirited bus

$\mathcal{G}_{enes}$ $\mathcal{P}_{roteins}$ $\mathcal{L}_{ipids}$ $\mathcal{S}_{ugars}$ $\mathcal{M}_{etals}$

| genome | proteome | lipidome | glycome | **metallome** |
| genomics | proteomics | lipidomics | glycomics | **metallomics** |

Figure 2.1 'omes'. The field of biometal sciences has matured to the point where it deserves a position next to the other 'omics' sciences that address the role of biomolecules at a systems biology level.

excursion to Mount Fuji in 2002, he envisioned metallomics as an integrative biometal science (Haraguchi 2004) with the following definitions (Lobinski *et al.* 2010):

"**Metallome**: Entirety of metal- and metalloid species present in a biological system, defined as to their identity and/or quantity".

The following notes were added:

"1. The metallome can be determined in a bulk biological sample representative of the system [or its component(s)] or at specific location(s).
2. The metallome can be characterized with different degrees of approximation, such as a set of (i) total element concentrations; (ii) metal complexes with a given class of ligands, e.g. proteins or metabolites, or (iii) all the species of a given element, e.g. the copper metallome.
3. In contrast to the genome of which the analysis has a specific endpoint (…) the description of a metallome … can never be complete. In particular, the numerous … metal complexes with biological ligands can be described only in terms of kinetic constants and defined thermodynamic equilibria.

   Metallomics: Study of the metallome, interactions, and functional connections of metal ions and other metal species with genes, proteins, metabolites, and other biomolecules in biological systems. A metallomics study is expected to imply: (i) a focus on metals (…) or metalloids (…) in a biological context, (ii) a link between the set of element concentrations or element speciation with the genome. This link may be statistical (…), structural (…) or functional (…); or (iii) a systematic or comprehensive approach. The identification of a single metal species, however important, without specifying its significance and contribution to a system should not be referred to as metallomics".

Hiroki Haraguchi also pointed out that in addition to the metallomes in the biosphere different metallomes can be investigated in the environment of organisms, e.g. in water (hydrosphere), soil (lithosphere), and air (atmosphere). Clearly metallomics can address metallomes in these spheres independent of biology. However, in this book the terms metallomes and metallomics will be employed only in relation to living organisms and their ecology.

Metallomes depend on multiple variables and are dynamic, and therefore they are snapshots in a multidimensional space, and as stated above "can never be complete". In order to describe the distribution of metal ions in a system, not only do we need information such as kinetics, thermodynamics in addition to analytical information about quantities and species, but also we deal with an infinite number of metallomes in a given species as a function of time and other variables such as genetic variability (variomics). The word *metametallomics* then refers to the field of metallomics as a function of such variables. An example of metametallomes would be the cancer metallomes that refer to the changes of metals occurring in different cancers or at different stages of cancer.

2.2 Structural, functional, and quantitative metallomics

Structural metallomics addresses metal coordination environments. Here, most of our knowledge is based on metal sites in proteins. Some knowledge exists about low molecular weight (LMW) complexes, but very little is known about the significance of the interaction of metal ions with other biomolecules with the exception of a role of magnesium in the structure of certain ribonucleic acids (RNA). Hence, proteins and LMW complexes feature prominently in this book.

Functional metallomics aims at understanding the functions of metal ions. Regarding proteins, functions are classified as catalytic, structural, and regulatory. Enzyme classification includes six categories: oxidoreductases, transferases, hydrolases, lyases, isomerases, and ligases. Use of metal ions in these categories varies, with redox-active metal ions featuring prominently in oxidoreductases, but other metal ions such as redox-inert zinc being present in enzymes of all classes. Function also varies. A catalytic metal ion can activate a bound water molecule, a substrate or a metal ligand, and it can participate in electron transfer. Structural metal sites can hold large parts of a protein together, organize domains of a protein or employ ligands from subunits of the same protein or different proteins to facilitate protein–protein interactions. In regulatory sites, metal ions are effectors of protein function, and usually bind transiently. Categories are not absolute as the function is not necessarily just one. Catalytic metal ions can organize the structure of the protein and hence have a structural function as well. Catalytic metal centers in some metalloproteinases, which are generally zinc metalloenzymes, exist in an inactive form (zymogen) with an additional protein ligand bound. Thus the site resembles a structural site. Dissociation of the additional protein ligand then gives way to water as a ligand, thus generating the site of the active enzyme (Figure 2.2).

The organization of protein domains by metal ion binding can be dynamic and result in a transition from a regulatory to a structural function, and last but not least, regulatory ions may participate in catalysis as in binuclear sites with a second, co-catalytic metal ion. But there are also other sites where metal ions are transported, stored, or sensed. Because of this variability of function, it is not straightforward to derive function from the primary coordination environment of metal ions in proteins. In fact, metal complexes in proteins with the same set of ligands can serve different functions. Often the type of the protein domain to which the

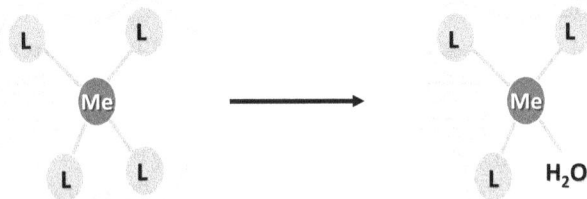

Figure 2.2 Conversion of a structural to a catalytic metal site. In some zymogens (precursors of proteinases), the protein keeps a catalytic site in an inactive state by providing an additional ligand to the metal (zinc). Dissociation of this ligand generates the catalytically active metal site.

metal ion binds assists in assigning function, but sometimes the wider context of the biological environment of the protein is necessary to define function.

Quantitative metallomics looks at the concentrations of the different species in the system. How the concentrations of all the metal species add up to the total metal concentration is a goal that has not been achieved for any metal ion. Concentrations of metal ions usually vary over many orders of magnitude. Since very small fractions of the total amounts of metal ions can be distributed among a large number of different species, quantification can account for the most abundant proteins but will require exquisitely high resolution methods to account for all low abundance binding sites. Investigations of metalloproteins also demonstrated the existence of abundant metalloproteins with unknown function in some species.

The information available in these three areas is far from complete. Complete annotation of structural, function, and quantitative metallomes is the goal of metallomics, though.

2.3 Metal speciation in biology: the metallomes

When we discuss metals in biology, we usually mean metal ions. Only in a few instances, the elemental metal is actually present. The term metal ion is straightforward for Na^+, K^+, Mg^{2+}, Ca^{2+} or Zn^{2+}, all of which occur in just one valence state in biology. Other metal ions exist in multiple valence states, requiring further specification, e.g. Fe(II), ferrous (Fe^{2+}), and Fe(III), ferric (Fe^{3+}), but also Fe(IV), the ferryl (Fe^{4+}) ion as a short-lived intermediate in enzymatic reactions. Transition metal ions may also occur in low and high spin states, referring to the distribution and pairing of electrons in different energy levels. The properties of the metal ions differ significantly in their valence and spin states. An important property is the exchange kinetics of ligands, which can vary from seconds, or even fractions of seconds to minutes and days. Co(III) and Cr(III) complexes are exceptional as they exchange their ligands extremely slowly. Another aspect of speciation is the interaction of metal ions with biomolecules, in particular with proteins. With increasing binding strength, the equilibrium between metal ion and macromolecular binding partner shifts to the right in Eq. (1) and the free metal ion concentration, i.e. the metal ions not bound to macromolecules, decreases.

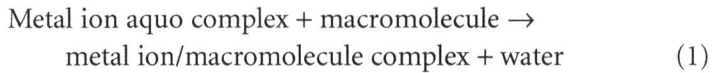

$$\text{Metal ion aquo complex} + \text{macromolecule} \rightarrow$$
$$\text{metal ion/macromolecule complex} + \text{water} \qquad (1)$$

In contrast to metal ions with low affinity for ligands, where the metal ion is primarily in the form of the aquo complex, e.g. Na^+, the higher affinity metal ions are present as ligand complexes, Eq. (2).

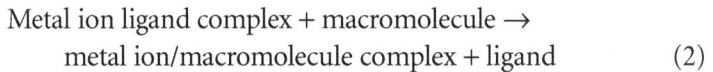

$$\text{Metal ion ligand complex} + \text{macromolecule} \rightarrow$$
$$\text{metal ion/macromolecule complex} + \text{ligand} \qquad (2)$$

Aside from water, the LMW ligands of biological metal ions are largely unknown.

Speciation describes the distribution of metal ions among the different chemical forms and the changes of these forms depending on ligand and metal concentrations, pH value, and redox potential. Particularly instructive for biology are so-called Pourbaix diagrams (Marcel Pourbaix, 1904–1998). These diagrams, here given for zinc, show the chemical species of metal ions as a function of both redox potential and pH (Figure 2.3). Notably, Pourbaix diagrams address only these two parameters. Additional speciation diagrams are required to indicate the different ligand complexes and their pH dependence.

Despite the fact that cells strive to maintain a constant pH and redox potential, it is the variation of both that is important for functions supporting life. The average redox potential outside human cells is more oxidizing compared to inside cells and there can be significant variations of redox potentials to make chemical reactions possible. With regard to pH, considerable variations occur when pH gradients are used in bioenergetics to synthesize ATP (pH value of the mitochondrial matrix about 8). In specific subcellular compartments (endosomes, lysosomes) pH values can be below 6. pH values change in parts of the digestive tract. The stomach is remarkable in having a pH value as low as 2. Clearly, these changes affect metal speciation, availability, and reactivity and therefore are critically important for metallobiochemistry. The diagram (Figure 2.3) shows

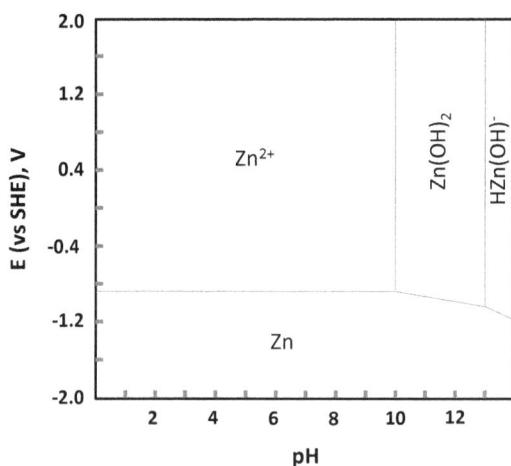

Figure 2.3 Pourbaix diagram for zinc. Zn^{2+} (in its aquo form) is stable over the ranges of redox potentials and pH values prevailing in biological cells and fluids. The diagram shows the amphoteric properties of zinc(II) ions, namely forming first an insoluble hydroxide at higher pH values and then, at even higher pH values, a soluble hydroxide in the form of $Zn(OH)_4^{2-}$ (ZnO_2^{2-}, not shown and not relevant to biology).

that zinc is always Zn^{2+} under almost all pH and redox conditions in biology. In contrast to zinc, transition metal ions occur in different redox states, which require special coordination environments for their stabilization, handling, and control. While global cellular pH and redox state are known, the cellular species of ions are often not known or specified correctly with regard to their localization in specific cellular compartments.

The main focus of metal biology has been the interaction of metal ions with proteins forming metalloproteins and therefore the coordination environments of metals in proteins are at the center of bioinorganic chemistry. The biological importance of the interaction of metal ions with other biomolecules is established only in a few cases. Exception proves the rule, though. The scholastics expressed it in Latin as "exceptio probat regulam", which means an exception tests the rule, and in this context that the significance of the interaction of other biomolecules such as RNA and DNA with metal ions has yet to be investigated in detail. Regarding biological ligands there is a major difference between biological inorganic chemistry and bioinorganic chemistry. Metal ions can bind to a variety of biomolecules and such interactions have been investigated extensively, but many interactions have no significance for biology. The reason is that the availability of metal ions and ligands is controlled at characteristic concentrations in biological systems thus limiting the types of interactions possible.

(a) LMW complexes

LMW compounds are involved in the extracellular acquisition of metals and in the binding of metal ions in the cell. They capture metal ions outside the cell when supply to the organism in the environment is limited. Many organisms have high affinity receptors for the binding and uptake of LMW metal complexes. For iron, such LMW complexes are known as sidero-phores. They are *chelating agents*. The name comes from the Greek (chele for the claw of a crab). The strength of the complex increases with the bite ("denticity") of a number of donors in a ligand. For example, ethylenediamine-tetraacetic acid (EDTA), a synthetic chelating agent, can coordinate metal ions with up to six donors (two nitrogens and four oxygens) (Figure 2.4A).

Using exceptionally high affinities, siderophores such as the *Bacillus anthracis* stealth siderophore (Figure 2.4B) ascertain that miniscule amounts of iron(III) available are tightly bound and the probability of losing iron is close to zero. More recently "phores" have been discovered for other metal ions: zincophores (Zn) based on ethylenediaminedisuc-cinic acid and chalkophores (Cu) such as methanobactin. The existence of nickelophores (Ni) has been postulated. Some antibiotics are also specific metal-binding compounds. Some of them are ionophores, which carry metal ions through biological membranes.

In addition, there are some intracellular chelating agents for metal ions, and here the tetrapyrroles feature dominantly. They are chelating agents par excellence and serve as ligands for at least four metal ions, iron, nickel, cobalt and magnesium (Figure 2.5A). They participate in fundamentally important reactions of life, such as oxygen transport in hemoglobin (iron) and energy harvesting in chlorophyll (magnesium). They are used as *cofactors* and *prosthetic groups*. These two terms are often employed synonymously though the term prosthetic group has been introduced for covalently bound, non-dissociable compounds.

Another chelating agent is based on a pterin structure and used for binding molybdenum (Figure 2.5B). LMW ligands such as metabolites are

Petrobactin

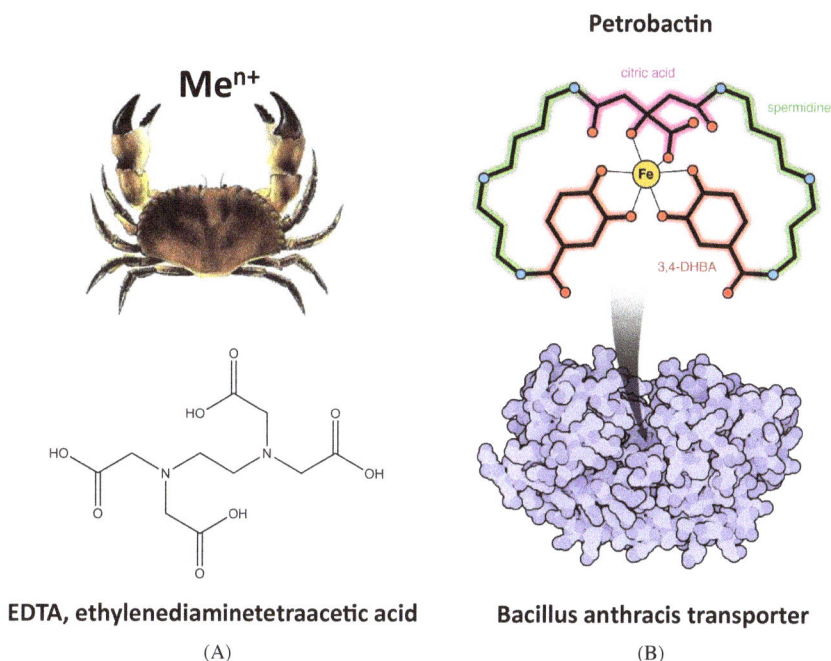

Me^{n+}

EDTA, ethylenediaminetetraacetic acid

(A)

Bacillus anthracis transporter

(B)

Figure 2.4 LMW ligands and their metal complexes. (A) The word "chelating" derives from the Greek word "chele" for the claws of a crab (image modified from ht). In a chelating agent, the metal ion is bound by at least two donor atoms in proximity and covalently linked in the ligand. EDTA is an example of a synthetic chelating agent. It binds metal ions with two nitrogen and four oxygen donors. (B) Siderophores are biological chelating agents for Fe^{3+}. Petrobactin, a siderophore from *Bacillus anthracis*, is composed of citric acid, spermidine, and 3,4-dihydroxybenzoic acid (DHBA). It binds to a bacterial receptor for uptake of iron by the bacterial cell but it evades human receptors and therefore is called a stealth siderophore. From the Structural Biology Knowledgebase; http://sbkb.org/articles/images/featuredsystem/petrobactin_3gfv.jpg.

also used in in the binding, e.g. sulfide, homocitrate, or glutathione (Figure 2.6) in Fe–S clusters.

Glutathione has been discussed as the major ligand for Fe(II) in the labile iron pool (Hider and Kong 2013).

Examples of complex metal cofactors in proteins are the iron-sulfur (Fe–S) clusters (Figure 2.7A), the MnCa cluster in the oxygen evolving complex of photosynthesis (Figure 2.7B), and the the iron-molybdenum cofactor (FeMoCo) in nitrogenase (Figure 2.7C).

Fe: Heme (porphyrin)

Ni: F430 (hydrocorphin)

Co: Vitamin B$_{12}$ (corrin)

Mg: Chlorophyll (chlorin)

(A)

(B)

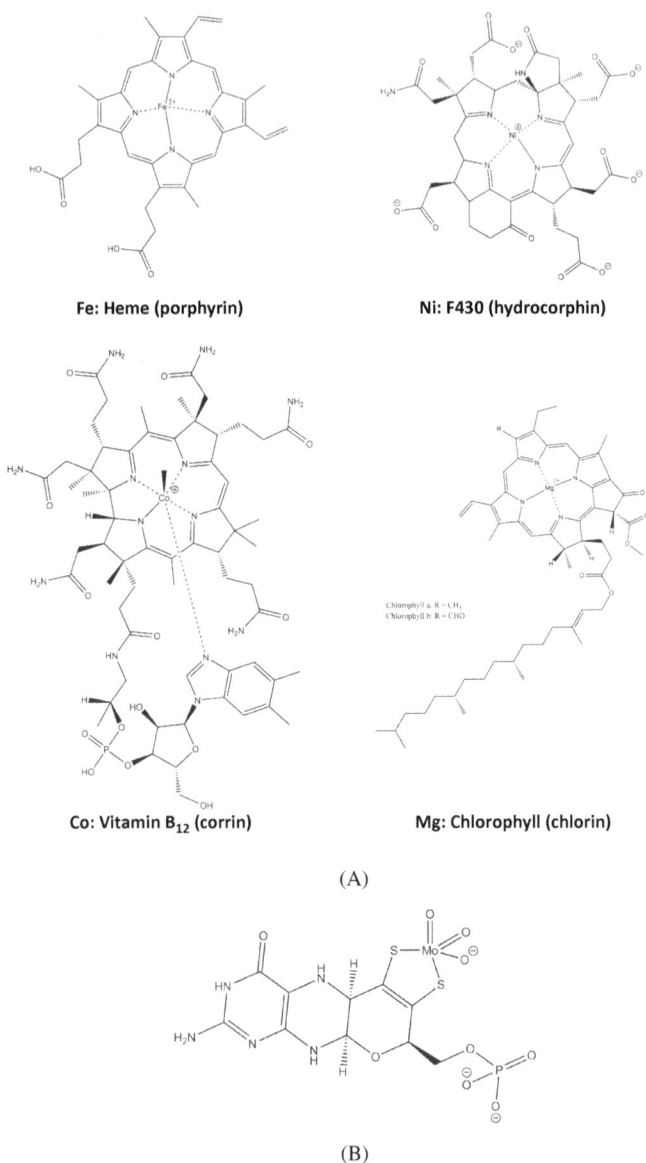

Figure 2.5   LMW ligands and their complexes. (A) Tetrapyrroles are biological chelating agents 'par excellence' using four nitrogen donors. They are employed in the metabolism of several metal ions: Fe in protoporphyrin IX (heme), Mg in chlorin (chlorophyll), Co in corrin (cobalamin), and Ni in hydrocorphin (F430); (B) The cofactor used for binding of Mo is based on a pterin structure. The donors of pterin are two sulfur atoms.

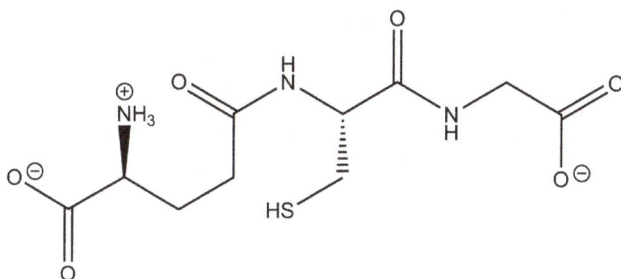

Figure 2.6 LMW ligand. The tripeptide γ-glutamylcysteinylglycine (glutathione, GSH) forms complexes with many metal ions. It can employ O, N, and S donors.

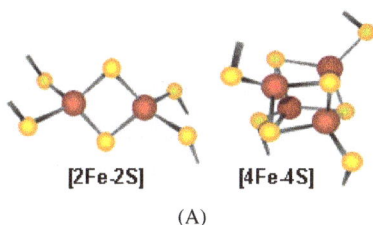

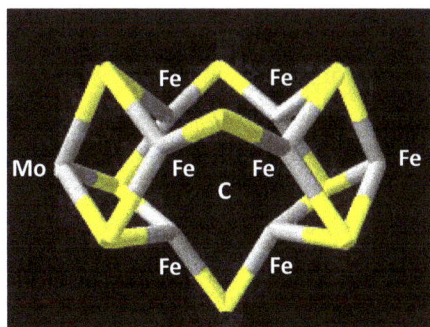

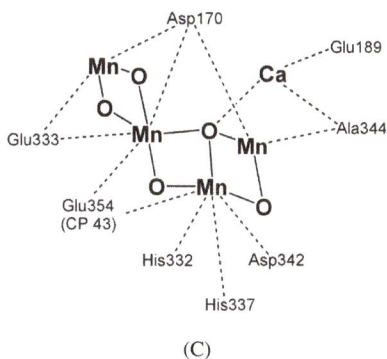

[2Fe-2S]   [4Fe-4S]

(A)

(B)

(C)

Figure 2.7 Examples of complex cofactors involved in redox reactions. (A) Fe–S clusters: 2Fe–2S and 4Fe–4S clusters (courtesy: Janneke Balk, University of East Anglia, Norwich, UK); the bridging sulfur is in the form of sulfide (S^{2-}); (B) FeMoCo of nitrogenases (modified from J.P. Juanita, Wikimedia Commons, released into the public domain); in some species, V substitutes for Mo in the cluster; the sulfur bridges are given in yellow; a C-atom is bound in the center of this cofactor; (C) the Mn, Ca cluster in the oxygen evolving complex (OEC) of plant photosynthesis; the bridging oxygen is in the form of oxide (O^{2-}) (from Yikrazuul, Wikimedia Commons, released into the public domain).

There are many more examples of di-, tri-, or tetra-nuclear arrangements of metal ions in proteins. These structures can form independent cofactors or form only when the protein folds. An example of the latter is mammalian metallothionein with its tri- and tetra-nuclear zinc clusters (Chapter 6, Figure 6.10A). A MgATP complex as the substrate for kinases is an example for the formation of *ternary complexes* Me.LMW compound. protein. In some instances, kinases seem to prefer a ZnATP complex.

Under physiological conditions, sugars and lipids do not seem to make functionally important contributions to transition metal speciation in cells. To which extent free amino acids (His, Cys, Glu, Asp) contribute as ligands of metal ions is not known. For those transition metal ions with high affinity and correspondingly very low concentrations of the free ion, the cellular ligands and complexes are not known. Cellular vacuoles have an environment different from the cytosol with a distinct metallobiochemistry. The ligands of vesicular zinc in subcellular compartments, for example, are not known, with the exception of the formation of an insulin/zinc complex in the granules of pancreatic β-cells.

(b) High molecular weight (HMW) complexes

The physiological significance of interactions of metals with RNA and DNA, except for the use of Mg in RNA molecules such as ribozymes — catalysts based on RNA rather than protein — is not clear. If cellular metal concentrations increase, they may target macromolecules such as RNA and DNA as is the case for metallodrugs interacting with DNA.

Given the functional significance of proteins, metal–protein interactions are the single most important area for the role of metals in biology. Therefore metalloproteomes and metalloproteomics take center stage in metallomics. The structural flexibility of proteins provides a large number of different geometries and coordination environments. Proteins either bind the metal ions directly as a cofactor or bind a prosthetic group with a bound metal. A limited number of amino acids have side chains with suitable donors for metal coordination. The donors are oxygens of Asp, Glu (in the case of iron also Tyr), nitrogens of His, and the sulfur of Cys (in the case of copper also Met) and in special cases the selenium of selenocysteine (Sec) or the oxygen of the carboxy group of γ-carboxy glutamic acid (Gla) in the case of calcium (Figure 2.8). The choice of ligands differs somewhat between the metal ions.

$$Me^{n+}$$

N	**O**	**S**
His	Glu, Asp, Tyr (Fe)	Cys, Met (Cu)

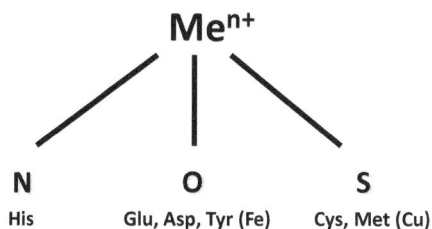

Figure 2.8 The amino acids that provide side chains with donors for metal binding in proteins. The donors (O, N, S) of metal ligands are primarily from the side chains of the amino acids histidine (His), aspartate (Asp), glutamate (Glu), and cysteine (Cys). The side chains of methionine (Met), tyrosine (Tyr), and serine (Ser) are used for binding some metal ions, and occasionally donors from the peptide backbone or the terminal carboxy or amino group are used. The presentation is not meant to indicate that metal sites in proteins have only three ligands. Coordination involves either one type of ligand or a combination of different ligands.

The three donors (O, N, S) can form a great number of coordination environments because they may be used with only one type of donor or with any permutation of these donors.

Based on only three donors of the four amino acids (Cys, His, Asp, Glu) that account for the majority of metal interactions, the coordination environments of metals in metalloproteins may appear quite simple. However, the amino acids have different binding modes and they can engage with one metal ion or bridge two metal ions (Figure 2.9).

Amino acids in the secondary coordination sphere can participate in hydrogen bonding to the ligands and the protein structure can provide environments with physical characteristics that influence the properties of the metal sites. Mg and Ca use practically only oxygen, whereas transition metal ions and zinc also use nitrogen and sulfur. Interactions are described by the Pearson classification of hard and soft acids (metal ions) and hard and soft bases (ligands). Hard metal ions such as Mg^{2+} tend to coordinate with hard donors of ligands, e.g. oxygen, while softer metal ions such as Zn^{2+} prefer softer donors, e.g. sulfur. Within the first transition metal series, with increasing atomic number, the tendency to coordinate with the sulfur donor increases, i.e. the "thiophilicity" increases.

The coordination environments of proteins provide limited specificity for metal ions. Many sites can bind other metal ions *in vitro*. For example,

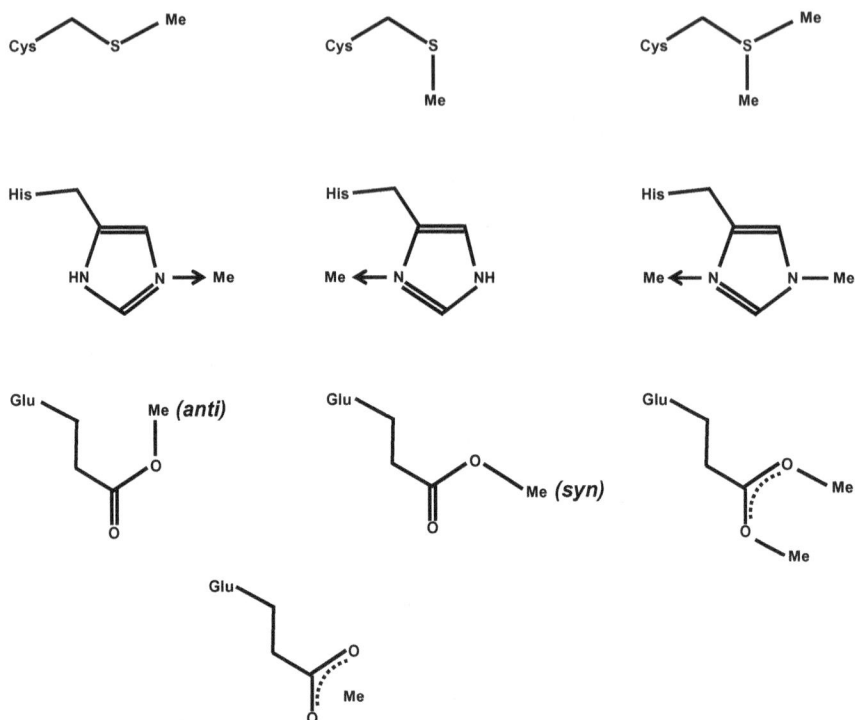

Figure 2.9 Different metal-binding modes of the donors of amino acid side chains in proteins. All three donors (sulfur from Cys; oxygen from Asp(Glu); and nitrogen from His) have the capacity to bind in different modes, including bridging metal ions.

liver alcohol dehydrogenase, a zinc enzyme with one histidine and two cysteine ligands in the active site, can bind copper, nickel, iron, cobalt and cadmium (not shown) instead of zinc, often while preserving catalytic activity (Figure 2.10).

Sites are somewhat optimized to handle the metal ions under the conditions prevailing *in vivo*, and yet, there is promiscuity in metal binding. Several metal ions can bind to the same proteins, e.g. metallothionein or *cambialistic* superoxide dismutases, which bind iron or manganese. How the correct metal is provided to a metalloenzyme is a matter of the involvement of other proteins (sensors) to be discussed later. A striking example of the specificity of metal incorporation *in vivo is* superoxide dismutase 1 (cytosolic) and 3 (extracellular), which need exactly one

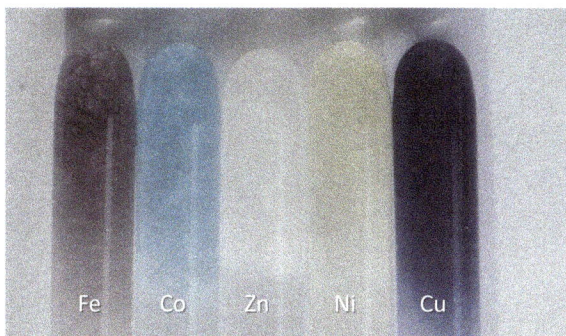

Figure 2.10 Metal-substituted alcohol dehydrogenase. The catalytic zinc ion in horse liver alcohol dehydrogenase can be removed specifically and the enzyme reconstituted with other divalent metal ions. The colour of the Fe(II)-reconstituted enzyme is due to the formation of Fe(III). The intense colour of the Cu(II)-reconstituted enzyme is due to the formation of a so-called type 1 copper site. The experiment demonstrates that *in vitro* enzymes accommodate many different metal ions in their active sites. *In vivo*, however, alcohol dehydrogenase is a zinc enzyme. From the author's personal archive.

copper and one zinc, and each metal ion is bound in the correct place of a binuclear site bridged by the imidazole ring of the side chain of histidine, despite the fact that cellular zinc and copper concentrations are quite different (Figure 2.11).

Coordination environments are different inside and outside a cell or in compartments within the cell. One factor is the difference in redox potentials. Due to the more oxidizing environment outside cells, metal ions exist in higher oxidation states and there are fewer free cysteines. Hence coordination to sulfur donors of cysteine is rare outside of the cell. The more reducing environment in the cell allows for a tendency to lower oxidation states and coordination to sulfur donors of cysteine. This principle is well illustrated by the use of Met (thioether) as a copper ligand outside and Cys (thiolate) inside the cell.

2.4 Metalloproteins

The functions that metal sites impart on proteins are structural, catalytic, and regulatory. In addition, proteins function in metal metabolism. In this case, it is the protein that imparts a function on the metal.

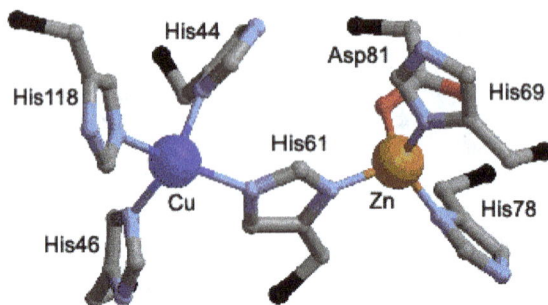

Figure 2.11   The active site of cytoplasmic superoxide dismutase. The metal-binding site is binuclear, requiring one copper ion and one zinc ion. Copper is the catalytic metal ion. A characteristic feature is a bridging histidine (imidazolyl bridge) between the metal ions. From the crystal structure of the bovine enzyme; Creative Commons Attribution Licence: J. Osredkar and N. Sustar (2011) *J. Clinic. Toxicol.* **S3:001.**

Metal–protein interactions have been classified into *metalloproteins* and *metal-activated proteins* on the basis of the affinities of the isolated proteins for metal ions: Metalloproteins are isolated with the metal remaining bound, while metal-activated proteins lose their metal during isolation (Vallee and Wacker 1970). This operational definition simply reflects the difference between strong and weak binding metal ions and does not consider the control of metal ions *in vivo*. In other words, according to this definition weak binding magnesium would always be involved in metal-activated proteins whereas strong binding zinc would always be involved in metalloproteins. Experience tells us that this is not necessarily so. The definition therefore requires a closer look as it would seem not to allow for regulation with strong binders or for forming metalloproteins with weaker binders. How can this paradox then be resolved? A definition based on isolation is arbitrary as the isolation is not performed at the "free" metal ion concentrations prevailing in cells. Metalloproteins with weaker binders such as magnesium, iron(II), or manganese(II) lose their metal unless sufficiently high free metal ion concentrations are present to saturate the binding site(s), which seems to be the case *in vivo*. Also, functional proteins are sometimes isolated in the presence of a chelating agent when inhibitory metal ions are present, resulting in the loss of the inhibitory metal ion. Again, this is an artefact of the isolation procedure. Some proteins bind an inhibitory zinc ion almost as tightly as genuine zinc metalloproteins bind zinc but since their isolation is based on enzymatic activity, conditions for isolation are such that they are kept active and that means the proteins are isolated in the presence of a chelating agent. This procedure masks the true behavior of the protein *in vivo*.

Apoproteins are made on the ribosome, the protein production facility in the cell, and bind the metal to form the holoproteins, the active metalloenzymes (Figure 2.12A).

However, metal ions also interact with proteins that are not recognized as metalloproteins and inhibit their action and then *the removal of the metal ion* generates the active enzyme (Figure 2.12B). Activation and inhibition of some proteins may occur in a very narrow range. Thus, a different definition than the one given above for metal-activated proteins and metalloproteins is required. It needs to include metal-activated as well as metal-inhibited enzymes. It is to be based on the equilibria between metal bound to proteins and metals not bound to proteins in the cell. Metalloproteins with low affinity for metal ions need high free metal ion concentrations. To describe them as metal-activated is an *in vitro* property that may not be relevant *in vivo*.

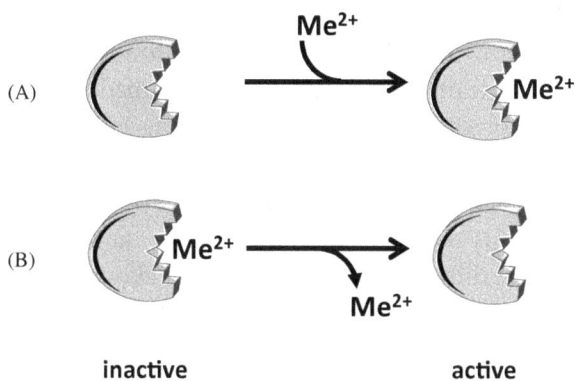

Figure 2.12 A,B Metal interactions with enzymes. (A) Inactive apoenzymes bind a metal (or metals) and become active holoenzymes. The site occupancy is usually 100% if the organism has no metal deficiency. (B) Alternatively, metals inhibit enzymes and these metal-inhibited enzymes are then activated by metal removal. In this case, occupancy may not be 100% if the metal participates in regulation of activity.

While this discussion concerns primarily metals controlling protein functions, there are yet other proteins where the sole function of the protein is the control of metal ions. Similar to regulatory metal sites, the metal is not present permanently but must come on and off while these proteins participate in moving metals to the location where they are needed. The above definition between metal-activated proteins and metalloproteins does not include these proteins. The definition of a metalloprotein therefore is even more complex with regard to proteins with tightly bound metal ions *and yet* participation in metal redistribution. Additional mechanisms for mobilization of metal ions exist in proteins that control metals. They have dynamic coordination environments that allow for binding and release and thus metal movement.

Metalloproteins where the metal ion imparts a function to the protein usually have integer stoichiometries. Since they depend on metal ions for their function it would be a waste of metabolic energy to synthesize a relatively large protein and then to compromise its function by not having enough metal available — assuming no metal deficiency in the organisms. However the proteins involved in metal redistribution and sensing may not have integer stoichiometries, challenging another tenet of metalloprotein biochemistry, namely that metalloproteins must have integer stoichiometries. One example is the extracellular Fe^{3+} transport protein transferrin. Its two metal binding sites are usually not saturated with iron. Despite its very high affinity for iron(III) ($K_a \approx 10^{23}$ M^{-1}) iron is released intracellularly after uptake of transferrin through receptor-mediated endocytosis in a process that involves acidification and binding of a synergistic anion (hydrogen carbonate). Another example is the intracellular protein metallothionein. It participates in cellular zinc buffering and transport, has relatively high affinity for Zn^{2+} ($K_a \approx 10^{11}$ M^{-1} for the strongest zinc binding sites), and is not fully saturated with zinc. Metallothionein releases zinc ions via oxidation of the sulfur donors of its cysteine ligands.

Many other proteins participate in metal movement: The regulatory protein calmodulin binds calcium, changes its conformation and then activates other proteins; membrane metal transporters move metal ions through biological membranes; metallochaperones safeguard metal ions and insert them into other proteins by binding to them and exchanging ligands; insertases and chelatases insert metal ions in prosthetic groups such as the tetrapyrroles; and then there are proteins that sense or store metal ions. For iron, the storage protein is ferritin. Iron(II) ions migrate on its protein backbone, are oxidized at a specific oxidase site to Fe(III) and then are deposited as iron oxide in its core, an inorganic "nanocage".

2.5 Metalloproteomes and metalloproteomics

After this consideration what constitutes a metalloprotein, the discussion turns to metalloproteomics, which aims at characterizing all the proteins containing a metal ion: the metalloproteome. In addition to analytical, experimental approaches, quite powerful predictive bioninformatics methods have been developed to identify metal binding sites in proteins. Recurring motifs (*signatures*) of amino acids that provide the ligand donors for metal binding, separated by a characteristic number of amino acids ("spacers"), were recognized in the sequence of metalloproteins when their three dimensional (3D) structures became available (Figure 2.13).

In some instances, recurring motifs were recognized in protein sequences and postulated to bind metal ions before 3D structures were known (Figure 2.14).

Zinc binding to these particular sequences (Figure 2.14) was confirmed and led to the identification of zinc finger proteins. The name for this metal-binding protein domain derives from the fact that zinc fingers bind to DNA in a way fingers would grab a rod (Figure 2.15A). Another protein domain is the calcium binding EF hand (Figure 2.15B).

Typical motifs for zinc, iron, and copper binding sites have been identified despite the fact that neither the spacers nor the ligands indicate a clear preference for the metal ion (Figure 2.16).

Metal-binding motifs can serve as templates for interrogating sequences of proteins for which 3D structures and metal binding sites are not known. Importantly, when DNA databases became available, nucleic acid sequences could be interrogated for codons of the amino acids that are part of such motifs. Thus, new zinc binding sites in enzymes and other proteins (Vallee and Auld 1990), calcium-binding domains, e.g. EF hands that obtained their name from the arrangement of helices in the protein, and zinc-binding domains such as zinc fingers, and other metal binding domains were identified by predictions, which turned out to have significant heuristic

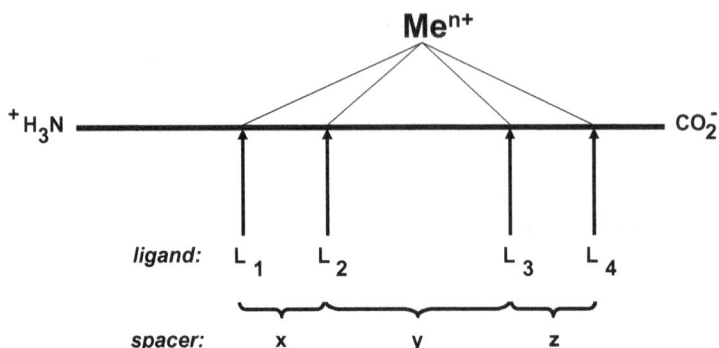

Figure 2.13 Metal-binding motifs in proteins. Ligand *signatures* are recognizable in the primary structure (sequence) of metal-binding proteins. Characteristic spacers (x, y, z), i.e. the number of amino acids between the ligands (L_1 to L_4) that provide the donor atoms, and the type of ligands determine the signature. While the coordination is shown for four ligands, metal-binding sites can have fewer or more than four ligands.

```
1   PVVYKRYICSFADCGAAYNKNWKLQ-AHLC-KH
2   TGEK-PFPCKEEGCEKGFTSLHHLT-RHSL-TH
3   TGEK-NFTCDSDGCDLRFTTKANMK-KHFNRFH
4   NIKICVYVCHFENCGKAFKKHNQLK-VHQF-SH
5   TQQL-PYECPHEGCDKRFSLPSRLK-RHEK-VH
6   AG----YPCKKDDSCSFVGKTWTLYLKHVAECH
7   QD---LAVC--DVCNRKFRHKDYLR-DHQK-TH
8   EKERTVYLCPRDGCDRSYTTAFNLR-SHIQSFH
9   EEQR-PFVCEHAGCGKCFAMKKSLE-RHSV-VH
```

Figure 2.14 Zinc finger sequence. Nine recurring signatures of cysteines and histidines in the sequence of the transcription factor TFIIIA from *Xenopus laevis* suggested a metal-binding motif (J. Miller, A.D. McLachlan, and A. Klug (1985) *EMBO J.* 4, 1609). They turned out to be a zinc-binding motif that became known as a zinc finger. The name is based on the analogy between the way this zinc-binding domain binds DNA or RNA and fingers grabbing a rod.

value. Once entire genomes of organisms became known, database could be mined on a larger scale. Through such searches for metal binding motifs it became possible to estimate the number of zinc, copper, and non-heme metalloproteins, i.e. the metalloproteomes, in entire organisms (Andreini *et al.* 2009). It led to the realization that the impact of metal ions

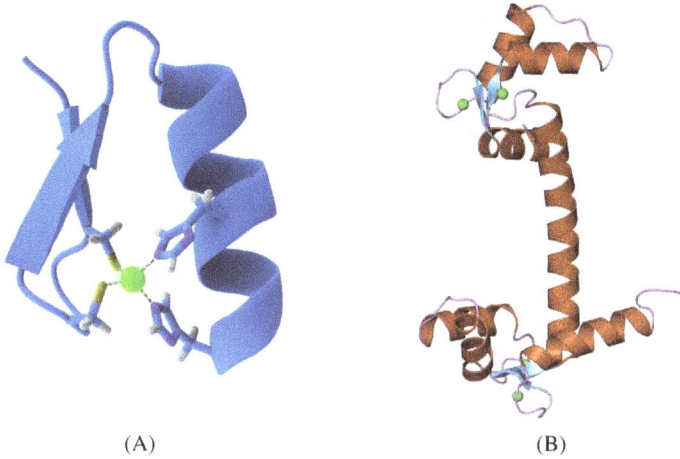

(A)   (B)

Figure 2.15   Structures of metal-binding domains. (A) The 3D structure of a zinc finger domain (Cys$_2$His$_2$ coordination) with a ββα secondary structure (from Thomas Splettstoesser, www.scistyle.com. Wikimedia Commons, under the GNU free documentation license); (B) The 3D structure of the four calcium EF hands in calmodulin from *Paramecium tetraurelia* (from http://www.ebi.ac.uk, Wikimedia Commons, released into the public domain). The analogy with a hand is based on the orientation of the "E" and "F" helices in the way a thumb and an index finger point.

Metal	Signature	Example
Fe	Cx$_2$Cx$_{29}$Cx$_2$C	rubredoxin
Cu	GMxCx$_2$C	copper-transporting ATPases
Zn	Cx$_2$Cx$_2$Cx$_7$C	structural zinc site in alcohol dehydrogenases

Figure 2.16   Examples of signatures for different metal ions. Metal-binding motifs for Fe, Cu, and Zn sites in proteins. The similar signatures of the iron-binding site of rubredoxin and the structural zinc-binding site of alcohol dehydrogenase illustrate that the type of metal cannot be predicted with absolute certainty from the signature alone.

on protein structure and function is much larger than was previously known and that about 30–40% of all proteins are metalloproteins. This realization alone, even without the functional implications for enzymatic catalysis and regulation, demonstrates how important metal ions are in cell biology. Based on isolated proteins, the previously estimated number of zinc proteins was in the order of a few hundred, but now with the new estimates based on whole genome databases the number turned out to be a few thousand zinc proteins. For the first time, the full impact of metal ions such as copper, iron, and zinc on biochemistry became evident, leading to the notion of a *galvanization of biology* in the case of zinc — an adage from Jeremy M. Berg. The numbers are remarkable. In eukaryota, about 9% of all proteins are zinc-binding proteins, about 1% non-heme iron proteins (in eurkarya), and <1% copper proteins. An estimate for manganese metalloproteins has not been attempted. The number of zinc proteins correlates linearly with the number of proteins in the 57 genomes analyzed. In prokaryota, the relative percentage of zinc proteins is lower and that of non-heme iron proteins higher. Additional methods, such as 3D information of metal binding domains and functional annotations of human gene sequences, improve the accuracy of predictions of metal sites in proteins further.

Noteworthy, estimates rest upon known signatures and assume that we know all signatures. They do not account for binding sites that are not readily recognized from motifs because the ligands are on different proteins when the metal ion bridges proteins or for sites where the spacers are so long that a motif is not readily recognizable, which is the case in some mono, bi-, tri-, and tetra-nuclear sites. Clearly a prediction is not an experimental analysis. It is often taken for granted that once a binding site is identified in a protein sequence, the metal must be there. Without direct analysis, this conclusion can be incorrect (Maret 2010). In some instances, the protein does not contain the predicted metal or it contains a different metal ion.

In contrast to these predictions of metalloproteomes, an experimental analysis of metal ions in a microorganism (*Paracoccus furiosus*) led to the remarkable finding that microbial metalloproteomes are largely uncharacterized and more diverse than previously thought (Cvetkovic *et al.* 2010). Instead of isolating proteins and characterizing their bound metal, metals were analyzed and then the associated proteins investigated. It was found that out of 343 discovered metalloproteins 158 were previously not known, thus identifying additional nickel and molybdoproteins and also proteins that have incorporated lead and uranium. Remarkably,

Thermolysin   VVAHELTHAVT...GAINEAISD
(*bacteria*)

LTA₄ hydrolase VIAHEISHSWT...FWLNEGHTV
(*human*)

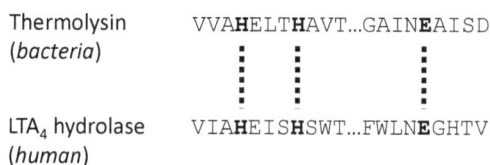

Figure 2.17   Prediction of metalloprotein function. Metal-binding sites in proteins *and their functions* can be derived from mining sequence databases for signatures. Based on sequence alignment, the known 3D (crystal) structure of the bacterial proteinase thermolysin allowed for the predictions that human leukotriene A$_4$ (LTA$_4$) hydrolase is a zinc enzyme and has proteinase activity (B.L. Vallee and D.S. Auld (1990) *Biochemistry* 29, 5647).

the organism accumulated the latter metal ions. A total of 21 metal ions out of 44 analyzed were found in the organism. This result demonstrates that metalloproteomes have not been fully characterized and that the importance of metal ions in protein structure and function is greater than the already remarkable estimates suggest.

Function also can be inferred by these predictive methods. For example, alignment of the sequence of human leukotriene A$_4$ (LTA$_4$) hydrolase with that of the bacterial proteinase thermolysin, which is known to be a zinc metalloproteinase, predicted that the hydrolase is a zinc enzyme with proteinase activity (Figure 2.17). Both predictions were confirmed experimentally.

Summary

Structural, functional, and quantitative metallomics focus on metal ions in biological systems. Metallomics is interdisciplinary and transdisciplinary: Analytical data are combined with biological assays and bioinformatics to understand how biometals function in biological systems. Metallomics includes speciation of metal ions bound in LMW complexes such as prosthetic groups and in macromolecules, among which metalloproteins constitute the most important class. Based on bioinformatics and experimental approaches, it has been estimated that 30–40% of all proteins contain a metal ion. The estimates of the metalloproteomes for individual metals such as iron, zinc, and copper leave no doubt about the importance of metal ions in biology. Metallomes apply not only to specific organisms but also to their ecology: air, water, and soil as parts of the environment in which life thrives. Catalytic, structural and regulatory functions of metals in proteins require coordination environments with specific characteristics. Metal sites in proteins have limited selectivity, which is determined by additional factors. A significant number of proteins function in the control and redistribution of metal ions with coordination dynamics in their binding sites. Despite powerful algorithms for prediction of metal sites in proteins our knowledge about metallomes is incomplete.

General reference

L. Banci, ed., Metallomics and the Cell, Metal Ions in Life Sciences, 12, Springer 2013.

Specific references

A. Cvetkovic, A.L. Menon, M.P. Thorgersen, J.W. Scott, F.L. Poole 2nd, F.E. Jenney Jr., W.A. Lancaster, J.L. Praissman, S. Shanmukh, B.J. Vaccaro, S.A. Trauger, E. Kalisiak, J.V. Apon, G. Siuzdak, S.M. Yannone, J.A. Tainer, and M.W.W. Adams (2010). Microbial metalloproteomes are largely uncharacterized. *Nature* 466, 779–784.

B.L. Vallee and W.E.C. Wacker, Metalloproteins, The Proteins, Vol. 5, H. Neurath, ed., Academic Press, New York 1970.

B.L. Vallee and D.S. Auld (1990). Zinc coordination, function, and structure of zinc enzymes and other proteins. *Biochemistry* 29, 5647–5659.

C. Andreini, I. Bertini, and A. Rosato (2009). Metalloproteomes: A bioinformatic approach. *Acc. Chem. Res.* 42, 1471–1479.

H. Haraguchi (2004). Metallomics as integrated biometal science. *JAAS* 19, 5–14.

R. Hider and X. Kong (2013). Iron speciation in the cytosol. An overview. *Dalton Trans.* 42, 3210–3229.

W. Maret (2010). Metalloproteomics, metalloproteomes, and the annotation of metalloproteins. *Metallomics* 2, 117–125.

R.J.P. Williams (2001). Chemical selection of elements by cells. *Coord. Chem. Rev.* 216, 583–595.

R. Lobinski, J.S. Becker, H. Haraguchi, and B. Sarkar (2010). Metallomics: Guidelines for terminology and critical evaluation of analytical chemistry approaches (IUPAC technical report). *Pure Appl. Chem.* 82, 493–504.

Chapter 3

The chemical elements of life

This chapter is a short excursion into some chemistry and biology in order to set a background and horizon for the systems that are addressed by metallomics. Importantly, it introduces the number of elements (metals and non-metals) essential for life and interacting with life. It explains the significance of metals for life and its evolution, and the interaction of metals with the environment. Last but not least, it gives some knowledge typically found in textbooks of bioinorganic chemistry, namely important biochemical reactions that depend on metals and the remarkable coordination environments of biometals.

3.1 Life and the elemental cycles

It is useful to keep in mind the great number of different organisms. Life is comprised of about 14 million species with estimates that about 90% of life forms are already extinct. A phylogenetic tree of life contains three domains: bacteria, archaea, and eukaryota (Figure 3.1).

Prokaryotes and eukaryotes have two drastically different cell types. The major difference is that eukaryotic cells have a nucleus, mitochondria, and organelles enclosed with membranes while prokaryotic cells do not have this subcellular structure. When we talk about the cell as the biological unit, we notice considerable variations. For instance, plants have chloroplasts in addition to mitochondria. The endosymbiosis theory holds that mitochondria and chloroplasts are remnants of bacterial cells once engulfed and incorporated into evolving eukaryotic cells. In the five kingdoms, eukaryotes are subdivided into protista, fungi, plants, and animals

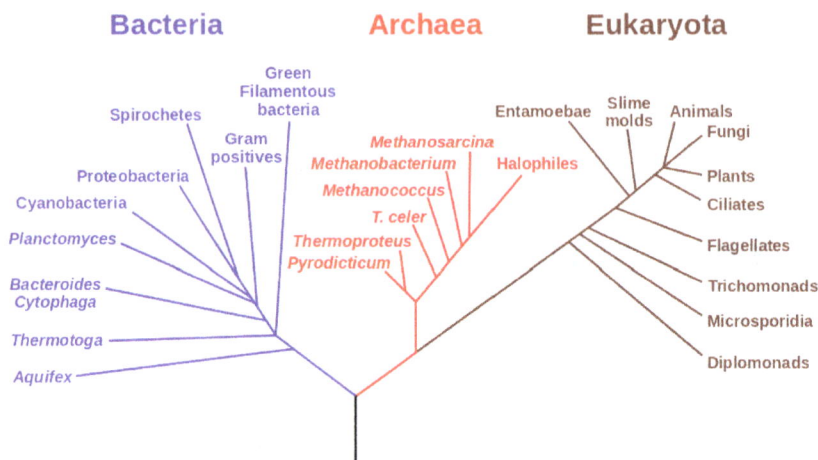

Figure 3.1   Phylogenetic tree of life. Biological species are presented in rooted evolutionary trees with three major branches (domains): Bacteria, Archaea and Eukaryota (from Wikimedia Commons, released into the public domain by its author Sting).

(Figure 3.2). It is important to keep this division in mind because there are significant differences in the way the different forms of life depend on metal ions. In fact, some metals are essential only for some forms of life, and there can be significant quantitative differences in the requirements for essential metals even in closely related species. Species differ in their genomes and consequently proteins. Differences among species can affect the way metal ions are handled or are functioning in subtle ways. Even in a given species such as humans there are genetic variations that affect metal metabolism.

Viruses are not part of the tree of life. They are considered a form of life although they do not fulfil all the criteria that define life and distinguish life from inorganic matter, namely the capacity for growth, reproduction, functional activity, and continual change preceding death (Oxford Dictionary). They also have considerable variability and include RNA and DNA viruses, with further subdivision into those having single stranded or double stranded nucleic acids. They also have proteins in their capsids. Viroids consist of single stranded RNA without a protein coat. Phages are viruses that infect bacteria. The replication cycles of viruses rely on metal ions from the host.

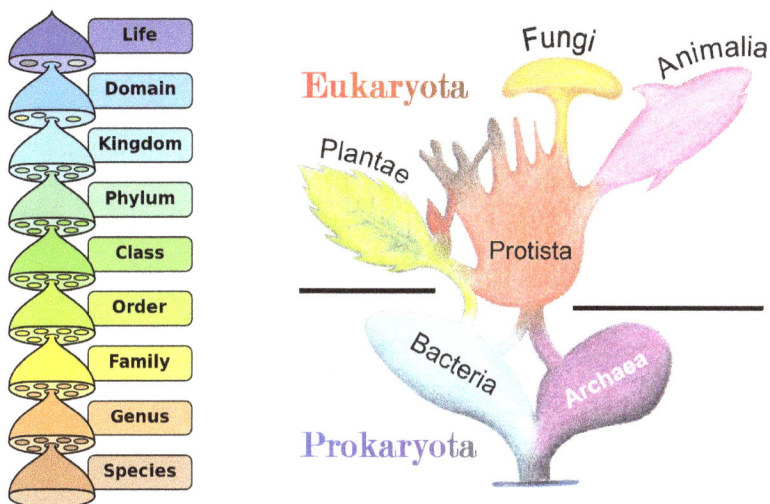

Figure 3.2   **Classification of Life. Left: Life is classified in eight taxonomic ranks. Right: The six kingdoms in the three domains of life, divided in eukaryota and prokaryota. From Wikimedia Commons released into the public domain by its author Peter Halasz (left) and Maulucioni y Doridí under Creative Commons Attribution-Share Alike 3.0 Unported license (right).**

In addition to symbiotic or parasitic relationship between organisms, organisms are active parts of their environments. Thus, geochemistry and biochemistry interact in the transfer of matter (Figure 3.3). Organisms participate in biogeochemical cycles of the elements that include the transfer of essential metals between the geosphere and the biosphere. They cooperate and in some cases the symbiosis of different forms of life supplies elements in the correct, useable chemical form. Many of the reactions in these elemental cycles are catalyzed by metalloenzymes with remarkable metal cofactors.

Organisms are classified according to the source of food (energy) they need for growth and survival. For carbon sources, they are called autotrophs when they rely solely on carbon dioxide (CO_2) and heterotrophs when they rely on reduced organic compounds.

The interrelationship of organisms is critical in the *ecology* of life. In the carbon cycle, plants introduce organic compounds into the world

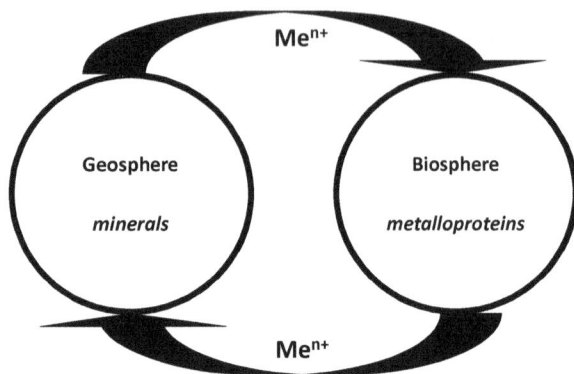

Figure 3.3 **Geobiochemistry. Microorganisms mobilize and deposit metal ions and determine the flow of metal ions between the geosphere (minerals) and the biosphere (metalloproteins).**

from CO_2, and we and other forms of life use these compounds and produce CO_2 (Figure 3.4).

Other reactions in the *carbon cycle* involve the gases carbon monoxide (CO) and methane (CH_4). Every step in the oxidation from the fully reduced compound (CH_4) to the fully oxidized compound (CO_2) involves a metalloenzyme:

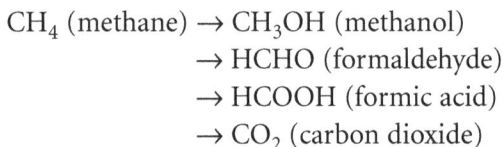

$$CH_4 \text{ (methane)} \rightarrow CH_3OH \text{ (methanol)}$$
$$\rightarrow HCHO \text{ (formaldehyde)}$$
$$\rightarrow HCOOH \text{ (formic acid)}$$
$$\rightarrow CO_2 \text{ (carbon dioxide)}$$

1. Methane monooxygenase: Cu
2. Methanol dehydrogenase: Generally use a pyrroloquinoline quinone (PQQ) cofactor; some enzymes employ lanthanides more efficiently than Ca
3. Formaldehyde dehydrogenase: Zn
4. Formate dehydrogenase: Mo

The hydration of CO_2 to form bicarbonate (HCO_3^-) is catalysed by the zinc enzyme carbonic anhydrase. It has the highest catalytic rate known for any enzyme. The uncatalyzed reaction is very slow and therefore this enzyme is necessary for gas exchange and pH buffering. In plants,

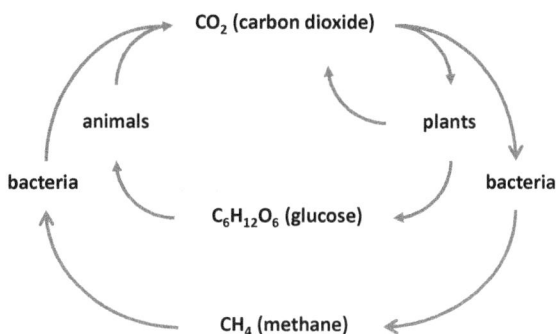

Figure 3.4   Carbon cycle. Central to the flow of carbon is the formation of carbon dioxide by oxidation of biological material by animals and plants and the reduction of carbon dioxide and hence its fixation in biomolecules (sugars) by plants. Fe is involved in the production of carbon dioxide while Mg is involved in the fixation of carbon dioxide. Methane is formed by methanogenic bacteria and oxidized by methanotrophic bacteria. The processes involve Ni, Co, Fe, and Zn.

CO_2 reduction in the Calvin cycle begins with the enzyme ribulose-1, 5-bisphosphate carboxylase (RUBISCO), which requires Mg for activity. Regarding energy sources, organisms are further divided into photo-trophs, which use light, and chemotrophs, which oxidize organic (organo-troph) or inorganic (lithotroph) compounds (Figure 3.5). Combinations of the terms are used, e.g. photolithotroph. This classification refers to the major source of food/energy and it does not imply that other compounds, such as essential metals, are not needed.

Due to their catalytic power, metalloenzymes also dominate other biogeochemical cycles. The carbon cycle is linked to the *oxygen cycle*, where photosynthesis produces oxygen and uses water and mitochondrial respi-ration uses oxygen and produces water (Figure 3.6).

The oxygen evolving complex (OEC) of plants produces oxygen from water and requires Mn and Ca while the terminal enzyme in the mito-chondrial respiratory chain, cytochrome c oxidase, produces water from oxygen and uses Fe and Cu. Several stages of electron transfer in the reduc-tion of oxygen (O_2, dioxygen) occur and additional enzymes are involved in controlling these species such as hydrogen peroxide (catalases and per-oxidases using Fe in cytochromes) and superoxide (superoxide dismutase using Cu and Zn or Fe/Mn) (Figure 3.6).

Prokaryota

Domain	Bacteria	Archaea			Eukaryota		
Kingdom	Eubacteria	Archaebacteria	Protista	Fungi	Plantae	Animalia
Energy	A/H	A/H	A/H	H	A	H

unicellular   mostly unicellular   multicellular

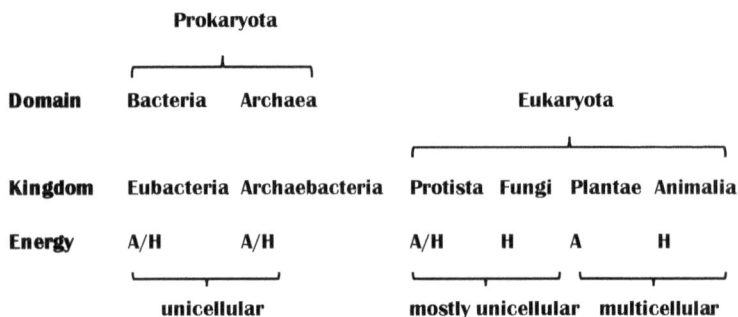

Figure 3.5   Further classification of life. The figure illustrates the cellular composition and the nutritional requirements in the different domains and kingdoms of life.

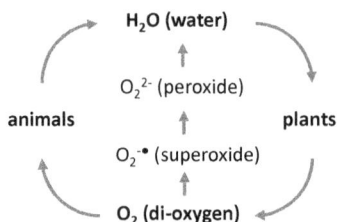

Figure 3.6   Oxygen cycle. Reduction of oxygen (O_2, dioxygen) occurs in different steps with species that need to be tightly controlled in biology: the rather reactive superoxide and (hydrogen) peroxide anions, which are involved in cellular signalling. The fully reduced state of oxygen is found in water (H_2O). Fe, Zn, Mn, and Cu are involved in the formation of the radical anion superoxide (indicated by the dot after the negative charge sign) and its dismutation to dioxygen and peroxide, and Fe is involved in the reduction of peroxide. Water oxidation also occurs in multiple steps and leads to the formation of oxygen. Mn and Ca and light energy are required to oxidize water to oxygen in the chloroplasts of plants. Fe and Cu are involved in the mitochondrial respiratory chain, which reduces oxygen to water. Monooxygenases, dioxygenases, and oxidases incorporate oxygen into biomolecules (not shown).

In the *nitrogen cycle*, the fixation of dinitrogen and its reduction to form ammonia (NH_3) make nitrogen available to organisms (Figure 3.7). Nitrogen fixation is a striking example of the efficiency of biological chemistry as compared to synthetic chemistry. An iron-based catalyst, high pressure (150–250 bar) and high temperature (300–550°C) are needed to synthesize ammonia in industrial plants. Microorganisms, however, carry out this reaction at ambient temperatures and atmospheric pressure. Intriguingly, they also use iron in the enzyme nitrogenase.

Elemental cycles are connected and additional ones have been described. Another but rather complex cycle is the sulfur cycle. Elemental cycles can also be constructed for the metal ions themselves.

One specific area where metal ions are crucial for catalysis is the fixation, metabolism and release of gases (Table 3.1). Moreover, some gases have been adopted for signalling purposes in eukaryotic cells and metalloenzymes are involved in controlling signalling. These gases have been called

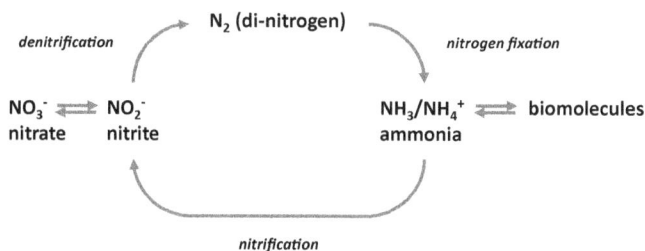

Figure 3.7 Nitrogen cycle. Reduction of nitrogen (N_2, dinitrogen) in nitrogen fixation results in the formation of ammonia (NH_3), which is the fully reduced state of nitrogen. Ammonia takes up a proton to form the ammonium ion (NH_4^+) in aqueous solutions. It is the starting point for incorporation (assimilation) of nitrogen into biomolecules. Dinitrogen fixation requires Fe and Mo or V. The oxidation of ammonia in nitrification produces nitrite (NO_2^-), which can be converted to nitrate (NO_3^-) in a reversible reaction requiring Mo. The reactions in nitrification employ Cu and Fe. Denitrification is the process in which nitrite is reduced to nitrogen with intermediates such as the gases N_2O and NO. The reactions also require Cu and Fe in metalloenzymes. Not shown is the anammox pathway of some bacteria, the anaerobic ammonia oxidation. In this pathway, enzymes such as iron metalloenzymes form nitrogen from ammonium and nitrite ions.

Table 3.1   Examples of metal-dependent reactions involving gases.

Hydrogen (H_2)

Hydrogenases, producing hydrogen; iron and nickel

$H_2 \leftrightharpoons 2e^- + 2H^+$

Oxygen (O_2)

Water oxidation to produce oxygen (in photosynthesis using light as energy); manganese and calcium

$2H_2O \rightarrow O_2 + 4e^- + 4H^+$

Oxygen reduction to produce water (in mitochondrial respiration); iron and copper

$O_2 + 4e^- + 4H^+ \rightarrow 2H_2O$

Nitrogen (N_2, N_2O, NO, NH_3)

Nitrogenases, reducing nitrogen to make ammonia, the most reduced form of nitrogen; iron and molybdenum (vanadium)

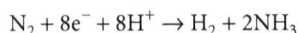

$N_2 + 8e^- + 8H^+ \rightarrow H_2 + 2NH_3$

Nitrite and nitrous oxide reductases producing nitrogen from nitrous oxide; copper

$N_2O + 2e^- + 2H^+ \rightarrow N_2 + H_2O$

Nitric oxide synthases producing nitrogen monoxide; iron, zinc, and calcium

L-arginine $+ 3/2NADPH + H^+ + 2O_2 \leftrightharpoons$ citrulline $+$ NO $+ 3/2NADP^+$

Carbon (CO_2, CO, CH_4)

Photosynthesis: reducing carbon dioxide to form glucose and oxygen; magnesium

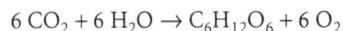

$6\ CO_2 + 6\ H_2O \rightarrow C_6H_{12}O_6 + 6\ O_2$

Mitochondrial respiration: oxidizing the reducing equivalents derived from glucose to from carbon dioxide and water; iron and copper

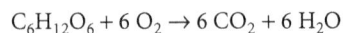

$C_6H_{12}O_6 + 6\ O_2 \rightarrow 6\ CO_2 + 6\ H_2O$

Methane monooxygenases, oxidizing methane, the most reduced form of carbon, to produce methanol; iron and copper

$CH_4 + O_2 + 2e^- + 2H^+ \rightarrow CH_3OH + H_2O$

Enzymes such as methyl-CoM reductase producing methane from carbon dioxide; nickel

$CO_2 + 8e^- + 8H^+ \rightarrow CH_4 + 2H_2O$

Carbon monoxide dehydrogenase producing carbon dioxide from carbon monoxide; molybdenum, copper, and nickel

$CO + H_2O \leftrightharpoons CO_2 + 2e^- + 2H^+$

Formate dehydrogenase producing carbon dioxide from formic acid; molybdenum and tungsten

$HCOO^- \rightarrow CO_2 + 2e^- + H^+$

Carbonic anhydrase forming bicarbonate (hydrogen carbonate) from carbon dioxide; zinc

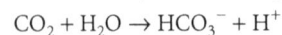

$CO_2 + H_2O \rightarrow HCO_3^- + H^+$

gasotransmitters. They include NO, CO, and H_2S. NO is produced from the amino acid L-arginine by nitric oxide synthases, which use Fe (heme) and Zn; CO is formed by heme oxygenase, and H_2S is formed by at least three enzymes.

3.2 Humans and bacteria: friend and foe

While emphasizing the diversity of organisms and differences in use of metal ions, both qualitatively and quantitatively, this book focuses on the interactions of metals with humans and bacteria and the relationship between the two. This focus has historic reasons because major insights into metallobiochemistry developed from investigations of bacteria.

Human activities are changing metal ion availability in the environment through industrialization, utilization of new materials, and new manufacturing processes. These activities cause nutritional deficiencies, in particular iron and zinc, but also increase the exposure to toxic elements by contaminating air, water, and soil (pollution) (Nriagu and Pacyna 1988). In contrast to some organic compounds, increased metal availability poses a particular threat to our health because metals are not biodegradable and can accumulate in trophic (food) chains and in organisms. Our body has to control the growth of commensal bacteria, i.e. its microbiome, and the virulence of invading pathogens. The trace metal status affects our susceptibility to infectious disease, the leading cause of morbidity and mortality worldwide and a major future threat because of the waning efficacy of available antibiotics due to bacterial resistance. Trace elements are important for our immune system and the ability to develop tolerance and resistance. They are required for both the host and the microorganisms. The competition for metal ions is a constant tug of war with elaborated molecular strategies of the microorganism to acquire metal ions for its metallome, e.g. the use of siderophores and other metallophores, and corresponding countermeasures of the host to deny such acquisition and preserve its metallome. The host employs antimicrobial proteins and peptides, some of which are metal-binding proteins. A major strategy of the host is nutritional immunity, a process in which concentrations of metal ions, in particular iron, manganese, and zinc are lowered in the blood so that they are no longer available for the microorganism. It happens in an acute phase response as the result of molecular signals from inflammation. It is a powerful defence against the invading organism. To defend itself against pathogenic microorganisms, the host also makes use of the cytotoxic properties of essential metal ions, e.g. copper. Likewise, it uses intoxication with reactive species, the formation of which involves the redox chemistry of metal ions. Metal deficiency is a protective adaptive response to limit metals available for microorganisms in regions of the world with a high prevalence of infectious disease. It is known for iron and may also pertain to zinc. In addition to extracellular mechanisms, the host defends itself intracellularly against parasites that have already invaded cells. Pathogens can destroy the tissue of the host to acquire the essential metal ions they need for growth or they can interfere with the host's homeostatic systems for controlling metal ions. Metals affect the innate and the adaptive immune response and immunopathology develops when changes of trace metal status causes inflammation. In contrast to disease resistance by the immune system, *tolerance* is a term that describes the capacity of the host to decrease damage to its tissues. In tolerance, the host is infected but remains asymptomatic. Metal ions may trigger autoimmunity and immunotoxicity. Toxicity of non-essential metal ions to the immune system also is an issue. Moreover, the toxicity of non-metal compounds increases when the organism is deficient of essential Zn, Fe, Se. Thus many essential and non-essential metals modulate the immune system at several levels. Clearly, these concepts apply to other species that live in a symbiotic relationship and they also contribute to the balance of species in a particular ecosystem.

3.3 The periodic system of the elements in biology

The Periodic Table accounts for all chemical elements and provides an order that reflects the physical and chemical properties of the elements (Figure 3.8). Dimitri Mendeleev (1834–1907) originally developed a table based on the periodicity of the chemical properties of the elements.

In the periodic system of the elements (PSE), the vertical groups/ blocks identify similar chemistry and the horizontal periods describe trends in properties. In addition, there is often a diagonal relationship of similar chemistries. Electronegativity, and hence the energy needed for ionization, increases from left to right and top to bottom, opposite to atomic radii. With the advent of understanding the electron configurations of atoms and ions, the PSE became also based on the physical properties of the elements. The order of the elements reflects the filling of s-, p-, d-, and f-electron shells and can be memorized by arranging the shells on a checkerboard (Figure 3.9).

The PSE has been extended with the discovery of additional elements and with those not present naturally and being laboratory-made only and radioactive (technetium and elements with atomic numbers $Z > 83$). It now includes 118 elements, thus almost filling the checkerboard (120 elements). Few of the 18 groups have specific names. Those having names are the alkali metals (group 1), alkaline earth metals (group 2), pnictogens (group 15), chalcogens (group 16), halogens (group 17), and noble gases (group 18). The elements in the two f-blocks are called the lanthan(o)ides and actin(o)ides. The important aspect for this book is the distinction among metals, metalloids, and non-metals; the metalloids are on a diagonal line separating metals and non-metals (Figure 3.10). Metalloids will not be treated in this book. The metals from group 3 to 12 are called transition metals. However, the chemists usually do not include group 12 in the transition elements. Hence, in this book we often refer to transition metals and zinc.

Non-metals function as ligands of metal ions. Selenium is a non-metal, a homologue of sulfur, though its element has a metallic *allotrope*. Elements differ by their number of protons. What is not directly seen in the PSE, but readily evident when comparing atomic numbers and masses, is the number of isotopes of each element. Isotopes have the same chemical properties (same number of protons) but differ in the number of neutrons and hence physical properties. For hydrogen, the two additional isotopes have different names: deuterium (one neutron) and radioactive tritium (two neutrons). The different physical properties make isotopes important for techniques of investigations. For example, several radioactive and stable isotopes of zinc are useful for investigations with different techniques. There are 25 radioactive isotopes of zinc, among which ^{65}Zn ($t_{1/2} = 244$ d) is most widely used, but others such as ^{62}Zn have

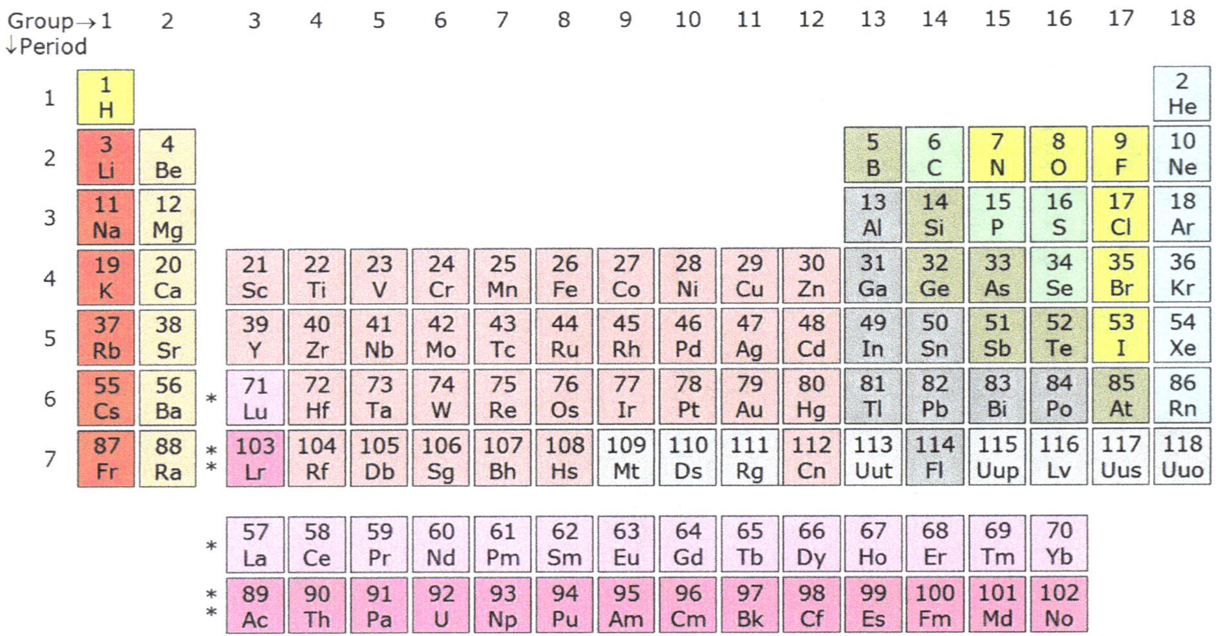

Figure 3.8   The periodic system of the elements (PSE). The elements are assigned to seven periods and 18 groups. Uuo for element 118 stands for Ununoctium (1-1-8). The lanthanides and actinides branch off after elements 56 (barium) and 88 (radium) and are usually presented below the PSE because inserting them would make the PSE too wide (from Wikimedia Commons, Creative Commons Attribution-Share Alike 4.0 International license). Color code: red: alkali metals; light brown: alkaline earth metals; light pink: lanthanides; dark pink: actinides; pale violet: transition metals; grey: post-transition metals; olive: metalloids; green: polyatomic nonmetals; light green: diatomic nonmetals; blue: noble gases; light grey: elements with unknown chemical properties.

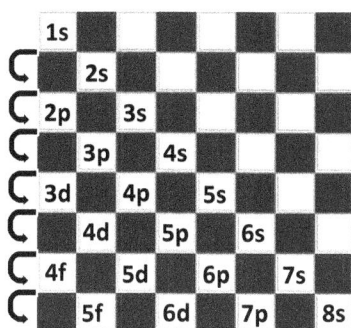

Figure 3.9 Electron configuration of chemical elements. The occupation of the different energy levels with electrons in shells and clouds, s, p, d, and f electrons, is the basis for building the PSE. A remarkable mnemonic for the building principle is the arrangement on a checkerboard.

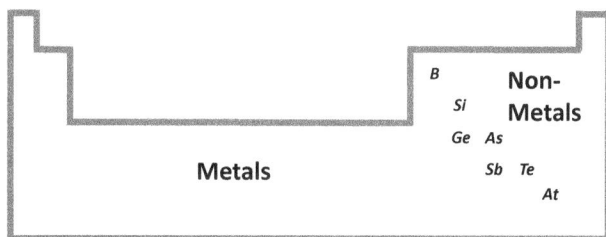

Figure 3.10 Position of metalloids in the PSE. Metalloids are positioned between metals (majority of elements) and non-metals (minority of elements) and have characteristics of both. The non-metals make up a triangle in the upper right part of the PSE.

been employed for proton emission tomography (PET), and there are five stable isotopes ^{64,66,67,68,70}Zn, which also have been used as tracers in biological experiments. Natural zinc contains 48.6% ^{64}Zn. Among the stable isotopes ^{67}Zn is used for nuclear magnetic resonance and Mössbauer spectroscopy.

Of interest for the subject matter of this book is a PSE that reflects the biological roles of metal ions, i.e. *a biological PSE*. The way such biological PSEs are usually presented generates considerable confusion due to at least three issues. First, the essential elements are given without reference to a specific form of life. Instead, what is presented is the sum of all elements essential for life despite the fact that the number of essential elements differs in organisms. Second, a clear definition is lacking what "essential" actually means, an issue that is discussed in the next chapter. Moreover, the relative biological significance of the elements is not apparent from such a presentation. Elements with critical general functions are mixed with those that have only very specific functions. Third and remarkably, it is often overlooked that many more metals and elements than just the essential ones are present in organisms and that their presence is often associated with functional outcomes (Table 3.2; Haraguchi *et al.* 2008; Roverso *et al.* 2015; Emsley 2001). Determining elements in biological material, of course, is a matter of sensitivity of the methods — an increasing number of elements can be found with increasing sensitivity. Organisms have the capability of metabolising, i.e. chemically modifying elements and their species, which is not evidence by itself for these elements having an essential function. For example, bacteria methylate mercury compounds and organisms synthesize arsenic-containing sugars and lipids. Such reactions are also part of the inorganic biochemistry of life.

It is instructive to collect the information about the abundance of elements in humans in the PSE (Figure 3.11). Nobel laureate Francis William Aston (1877–1945) determined the abundance of the elements on earth and found that it decreases with increasing atomic number. It is therefore not surprising to find that most of the abundant elements are employed in life with the notable exception of aluminium. On the other hand, some elements that are quite rare, e.g. selenium, have been acquired for essential functions.

Based on the quantities at which the elements occur in most biological matter, they are grouped into the bulk elements (in the order of total amounts): O, C, H, and N (g/kg), which are the main constituents of the four classes of biochemical building blocks: amino acids, sugars/carbohydrates, fats/lipids and nucleobases. This group also contains minerals at quantities that are a bit lower, but also in the g/kg range: Ca, P, K, S, Na, Mg, among which sulfur is also a component of building blocks, e.g. in the amino acids cysteine and methionine. Other groups of elements occur at mg/kg (parts per million, ppm) and μg/kg (parts per billion, ppb) levels. The concentrations of yet other elements are at the ng/kg (parts per trillion, ppt) level.

Table 3.2   Metal concentrations.

Metal	Salmon Eggs[a] ng/g	Seawater[a] ng/mL	Blood[a] ng/mL	Placenta (m)[b] ppm/dry wt.	Average Composition[c] g/70 kg	(ppm)
K	1860000	399000	151000	11000	1200	(17143)
Ca	432000	412000	93100	1300	125	(1786)
Na	247000	1078000	3130000	13000	100	(1429)
Mg	222000	1280000	17500	450	25	(357)
Zn	13600	0.35	651	52	2.3	(32.9)
Fe	10700	0.03	1200	480	4.0	(57.1)
Cu	8900	0.15	750	5.0	0.07	(1.0)
Sr	3600	7800	33	0.32	0.32	(4.6)
Ti	1420	0.0065	—	0.74	0.70	(10)
Rb	523	120	170	14	0.38	(9.7)
Mn	512	0.02	0.57	0.35	0.012	(0.17)
Ba	47.3	15	0.48	0.26	0.02	(0.29)
Ni	26.5	0.48	0.23	1.0	0.015	(0.21)
V	18.8	2.0	0.031	0.044	—	—
Co	12.5	0.0012	0.11	0.017	0.002	(0.03)
Hg	12.0	0.00014	0.55	0.0073(w)	—	—
Ag	10.7	0.002	0.2	—	—	—
Pd	8.0	0.00006	—	—	—	—
In	7.0	0.0001	—	—	—	—
Al	6.44	0.03	1.8	4.6	0.06	(0.86)

(*Continued*)

Table 3.2 *(Continued)*

Metal	Salmon Eggs[a] ng/g	Seawater[a] ng/mL	Blood[a] ng/mL	Placenta (m)[b] ppm/dry wt.	Average Composition[c] g/70 kg	(ppm)
Mo	6.43	10	0.95	0.021	0.005	(0.07)
Ge	5.83	0.005	—	—	—	—
Cs	5.58	0.31	0.95	0.035	—	—
Cd	1.1	0.07	0.15	0.041	0.02v	(0.29)
Sn	0.96	0.0005	0.51	0.036	0.03	(0.43)
Be	0.80	0.00021	0.09	—	—	—
Cr	0.73	0.21	0.069	0.13	0.002v	(0.03)
U	0.66	3.2	0.31	—	—	—
Pb	0.48	0.0027	1.2	0.065	0.12	(1.7)
Ru	0.19	0.000005	—	—	—	—
Sb	0.12	0.2	2.3	0.25	—	—
Au	0.054	0.0066	0.00003	—	—	—
Tl	0.050	0.0130	—	—	—	—
W	0.033	0.01	0.34	—	—	—
Pt	0.043	0.0002	0.0014	—	—	—
Ga	0.0023	0.0012	—	0.045	—	—

(a) Others not given are in the ppt range.
(b) Human placenta, m = maternal part median, i.e. × 1000 compared to first column, influenced by the content of blood; w = whole placenta, n = 19.
(c) From total ashing of human corpses, v = variable, converted to ppm for comparison.

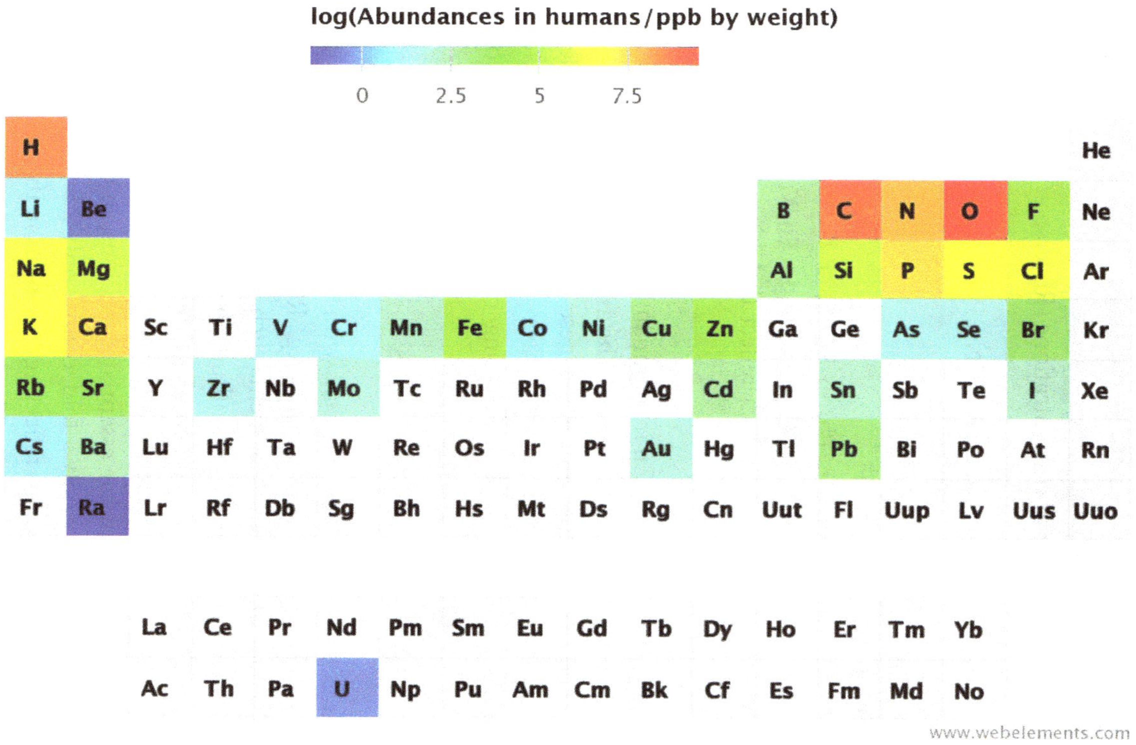

log(Abundances in humans / ppb by weight)

0   2.5   5   7.5

www.webelements.com

Figure 3.11 A PSE with the abundance of the chemical elements in humans. The bulk elements are in olive, brown, and red and they are all essential. In green and magenta are both essential and non-essential elements at similar concentrations. In blue are three elements with usually extremely low concentrations: Beryllium, uranium, and radium. Note that some additional elements are also present (Table 3.2) but not given in this Figure. In group 3, the last elements (Lu, lutetium and Lr, lawrencium) instead of the first elements of the f-block elements are given (compare with Figure 3.8). From www.webelements.com (Courtesy: Mark F. Winter, the University of Sheffield, reproduced with permission).

The low abundance elements are called trace or ultratrace elements. Originally, their quantification in tissues was challenging or impossible. With modern instrumentation, quantification is no longer a major issue and even levels below the ppt range can be measured. In human placenta, for example, the following elements were measured and found to cover at least nine orders of magnitude in concentrations (Roverso *et al.* 2015):

(a) >1000 ppm (dry weight): C, Ca, H, K, N, Na, P
(b) Around 100 ppm: Cl, Mg, Fe, S, Zn
(c) Around 10 ppm: Br, Rb
(d) Around 1 ppm: Al, Cu, Ni, Se, Ti
(e) Around 100 ppb: Ba, Cr, Ga, Mn, Pb, Sb, Sr, Ta, Zr
(f) Around 10 ppb or less: As, Cd, Co, Cs, Hg, Mo, Sc, Sn, V, and the lanthanides, the concentrations of which reflect their abundance in the environment.

The overall amounts of Fe and Zn in a human body of 70 kg are a few grams. Strictly speaking these are not traces. The amount of copper is one order of magnitude less (100 mg) and manganese (12–20 mg), molybdenum (5 mg), chromium (2 mg) and cobalt (2 mg) are yet one and two orders of magnitude less. Thus, in decreasing order, the amounts of d-block metals in humans are:

$$Fe \approx Zn > Cu > Mn > Mo > Co, Cr$$

Remarkable is the occurrence of metals with no known function in humans at relatively high concentrations, e.g. Rb, Ti, and Sr. Equally remarkable is the fact that essential elements such as Co and Mo occur at concentration that are 2–3 orders of magnitude lower than those of other essential elements and lower than some non-essential or toxic metals, e.g. Pb. The concentrations of some metals are rather constant whereas those of others vary considerably. For the essential metal ions, in the maternal part of the placenta, zinc varies only 3 fold, copper only 2.5 fold, molybdenum 4 fold, iron 15 fold (perhaps reflecting variations in the amount of blood in the preparations), manganese 20 fold, but cobalt 1200 fold. This variation indicates very tight control of the more abundant essential elements. The large variation in the case of cobalt does not indicate that there is no tight control of the vitamin B_{12} cofactor; it may indicate that the mechanisms to discriminate cobalt(II) when not bound in the cofactor from the other divalent metal cations are not very effective. Other elements vary due to variable exposure. The significance and functional consequences of large variations are not known. Bioaccumulation occurs for both essential and non-essential elements and therefore is not indicative of essentiality. It is high for the essential elements but also can be high for some non-essential elements, which, if not taken up for a purpose, may reflect inefficient discrimination. For salmon eggs vs. sea water, the bioaccumulation factor for iron is 0.5×10^6. Because of the extremely low solubility of Fe(III) ions at neutral pH in the absence of chelating agents, specific mechanisms for iron uptake are needed. Microorganisms use siderophores for the purpose. The bioaccumulation factors for copper, zinc, and manganese also exceed 1×10^4. Toxic mercury also has a high bioaccumulation factor. Silver, tin, gold, and germanium are accumulated with a factor of 1×10^3 though their overall concentrations in the egg cell are rather low. Remarkably, all the natural elements can be determined in a cell (74 out of 78 measured) (Haraguchi *et al.* 2008). Such analytical data demonstrate the importance of an inorganic perspective in biochemistry.

The metals Mn, Fe, Co, Cu, Zn, and Mo (given in their order in the PSE) are firmly established as being essential for many forms of life. To the essential non-metal trace elements Se and I, we can now add Br (McCall *et al.* 2014), thus defining a triangle of essential elements in the upper corner of the PSE, albeit with the understanding that fluorine is beneficial only for tooth health within a narrow range of intake. The issue what *essential* means becomes more complicated when discussing metals such as V, Cr, Ni, the metalloid As, and non-metals such as B, Si and F, where there are either very specific functions or where it remains unclear whether or not the elements are indeed essential for a particular form of life. B and Si, for example, are essential for some forms of life. B is the only essential element of group 13. All the other elements in this group (Al, Ga, In, Tl) are apparently not used, though some are present, e.g. there is 60 mg Al in the human body, i.e. five times more than essential Mn. Group 13 defines a vertical line of demarcation through the PSE separating essential metals from essential non-metals. Nickel (Ni) is essential for some organisms and has a firmly established role in a few of their metalloenzymes. Likewise, vanadium (V) has a role in nitrogenases and haloperoxidases and is present in marine organisms (tunicates), some algae, and fungi. Though Ni and V can be present in human tissues at concentrations higher than those of the essential metals Co and Mo, it remains unclear whether they have essential functions in humans. A special case is Cr. It had been firmly established as an essential trace element for humans and as such it is included in governmental guidelines for dietary advice. It is added to nutritional supplements such as multivitamin tablets, and some foods are fortified with it. However, its status as an essential element has been challenged recently. The arguments are involved and will be discussed in Chapter 4. While some elements such as F have very limited and special functions, other elements are restricted to specialized organisms. Cadmium occurs in carbonic anhydrase of some marine organisms (diatoms). Some organisms rely on tungsten (W) rather than molybdenum (Mo) as a cofactor for their enzymes. In contrast to the s-block and the d-block metals, apparently no p-block metal has been found to be essential — juxtaposing triangles of non-essential metals and essential non-metals in that block. The biological roles of the essential and non-essential elements are discussed in the following chapters.

Two types of biological PSEs are presented here (Figures 3.12 and 3.13). They will serve as the basis for the discussion in the next chapters. The first one (Figure 3.12) shows the essential metals and non-metals in humans. Noteworthy, chromium is left out in this PSE based on present controversies whether or not this metal is indeed essential for humans despite the fact that it had been rather firmly established as being essential. The second one (Figure 3.13) lists the essential metals and non-metals for a variety of organisms. It also leaves out chromium, and it leaves out elements that have either very specialized functions or are essential for very specific organisms in special environments.

Looking at the biological significance of the elements, there is a rather clear line of demarcation after the fourth period: In the upper part of the PSE most of the elements are believed to be essential (atomic numbers $Z < 36$) for some forms of life with the exception of Li, Be, Al, Sc, Ti, Ga, Ge, and, of course the noble gases, while almost all the elements in the lower part are believed to be non-essential with the exception of Mo and I and some special cases discussed later. The recognition that many more elements than the essential ones are present in organisms and that

H																	He
Li	Be											B	C	N	O	F	Ne
Na	Mg											Al	Si	P	S	Cl	Ar
K	Ca	Sc	Ti	V	Cr	Mn	Fe	Co	Ni	Cu	Zn	Ga	Ge	As	Se	Br	Kr
Rb	Sr	Y	Zr	Nb	Mo	Tc	Ru	Rh	Pd	Ag	Cd	In	Sn	Sb	Te	I	Xe
Cs	Ba	La*	Hf	Ta	W	Re	Os	Ir	Pt	Au	Hg	Tl	Pb	Bi	Po	At	Rn
Fr	Ra	Ac*															

Figure 3.12   The elements that are essential for humans. Chromium is given on a different background because of on-going discussions about its status as an essential element (see text). Minimally 11 metals and 10 non-metals, i.e. a total of 21 elements are essential.

H																	He
Li	Be											B	C	N	O	F	Ne
Na	Mg											Al	Si	P	S	Cl	Ar
K	Ca	Sc	Ti	V	Cr	Mn	Fe	Co	Ni	Cu	Zn	Ga	Ge	As	Se	Br	Kr
Rb	Sr	Y	Zr	Nb	Mo	Tc	Ru	Rh	Pd	Ag	Cd	In	Sn	Sb	Te	I	Xe
Cs	Ba	La*	Hf	Ta	W	Re	Os	Ir	Pt	Au	Hg	Tl	Pb	Bi	Po	At	Rn
Fr	Ra	Ac*															

Figure 3.13   The elements that are essential for different forms of life. Chromium is given on a different background for the reason discussed in the legend of Figure 3.12 and in the main text. In comparison to Figure 3.12, silicon, boron, nickel, vanadium, and tungsten have been added, bringing the total to 14 metals and 12 non-metals, i.e. 26 essential elements. Not considered are those elements that have been found in only rather unique organisms living in special ecological niches.

non-essential elements also have a biological chemistry is an important change in our thinking and a major point in the following discussions that aim at generating a broader understanding of how the elements in the PSE interact with biological systems and that suggest considering a *biochemis-*

try of the entire PSE to fully understand the role and impact of chemical elements in biology. In addition to the elements already present naturally there are those introduced by man-made activities. Some elements used in diagnosis and medical treatments are introduced into our bodies intentionally. Examples include silicon in breast implants or metals in joint prostheses, where the leaching of the elements into tissues can be a health issue. In these instances, and also in total parenteral nutrition, the normal barriers are by-passed and contamination and/or exposure can have serious effects. With the use of rare elements in the electronics industry and new materials such as nanoparticles, we are exposed to metals for which the biological effects are largely unknown. It is assumed that our bodies can somehow handle these elements, but there is no proof. In fact, on the basis of the mechanisms of handling metal ions discussed later, it is quite unlikely that we can cope with the exposure to certain metal ions. We are exposed to metals that express toxicity at low concentrations or are believed to be inert, and yet others, for which there is virtually no knowledge about biological effects. A view of the entire PSE and how the elements interact with biological systems supports the merits of a metallomics approach.

Summary

Organisms determine the flux of elements between the geosphere and the biosphere in biogeochemical cycles. In the majority of reactions in these cycles, metal centers catalyze particularly challenging inorganic reactions, such as the reduction and oxidation of gases. As catalysts, and through additional functions, metal ions were critically important for chemical and biological evolution, abiogenesis and biogenesis (Williams 2007). Metal ions also critically determine the symbiotic or parasitic relationship between organisms. A significant number of the chemical elements are essential for life, but not all them are needed for one particular form of life. Most of the elements in the upper half of the PSE are used for functions in biomolecules. An even larger number of elements are actually present in organisms and many of them participate in metabolism. The presence of some varies depending on exposure. There is limited knowledge about the functional consequences of the presence of non-essential elements but it is clear that many of the elements present are not chemically inert. Considering the PSE as a basis for biochemistry reduces the bias of biochemistry towards organic chemistry by including inorganic chemistry.

General references

J.O. Nriagu and E.P. Skaar, eds., Trace Metals and Infectious Disease, MIT Press 2015.

R.J.P. Williams (2007). A chemical systems approach to evolution. *Dalton Trans.* 10, 991–1001.

Specific references

A.S. McCall, C.F. Cummings, G. Bhave, R. Vanacore, A. Page-McCaw, and B.G. Hudson (2014). Bromine is an essential trace element for assembly of collagen IV scaffolds in tissue development and architecture. *Cell* 157, 1380–1392.

H. Haraguchi, A. Ishii, T. Hasegawa, H. Matsuura, and T. Umeura (2008). Metallomics study on all-elements analysis of salmon egg cells and fractionation analysis of metals in cell cytoplasm. *Pure Appl. Chem.* 80, 2595–2608.

J. Emsley, Nature's building blocks, University Press, Oxford 2001.

J.O. Nriagu and J.M. Pacyna (1988). Quantitative assessment of worldwide contamination of air, water and soils by trace metals. *Nature* 333, 134–139.

M. Roverso, C. Berté, V. Di Marco, A. Lapolla, D. Badocco, P. Pastore, S. Visentin, and E. Cosmi (2015). The metallome of the human placenta in gestational diabetes mellitus. *Metallomics* 7, 1146–1154.

Chapter 4

Essential metals

4.1 Dose-response curves for essential metal ions

The biological action spectrum of an essential nutrient such as a metal ion is described with three regions: decline of function at low concentrations, a plateau of optimal function at intermediate concentrations, and a decline of function at high concentrations. Decline of function is associated with disease, and with worsening of the condition it will eventually lead to death. The concept of such bell shaped curves was suggested by Gabriel Bertrand (1867–1962) and is known as Bertrand's rule (Figure 4.1).

The curves are used to estimate requirements and safe intakes for selected populations based on the definitions:

EAR, the Estimated Average Requirement, which is the 50th percentile for the requirement in the diet based on metabolic balance studies or other studies, and reflects the intake adjusted on assumptions of bioavailability.

RDA, the Recommended Dietary Allowance, which is the EAR plus two times the standard deviation and covers 97–98% of a population.

RDI, the Recommended Daily Intake.

NOAEL, the No Observable Adverse Effect Level, which is the level of exposure where no statistically significant adverse effect is observed in a population.

LOAEL, the Lowest Observable Adverse Effect Level.

UL, the Upper Level of intake, which is the highest tolerable level of intake which likely poses no adverse health effect for almost all individuals.

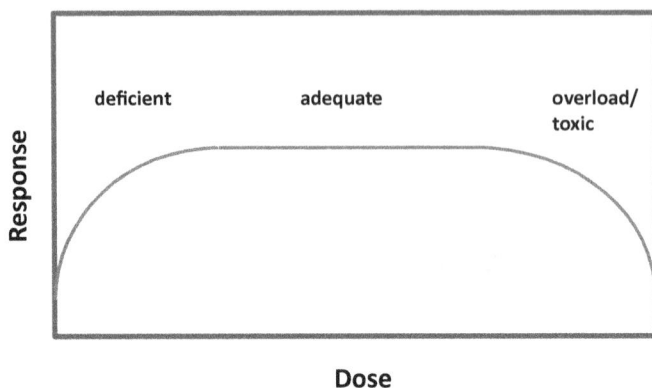

Figure 4.1   Dose/response curve for essential metal ions. The idealized curve has three regions: negative effects at low concentrations, a plateau with optimal functions at intermediate concentrations, which can be rather narrow, and negative effects at high concentrations. Not shown is a biphasic response that some metals may have: a pharmacological effect before they become toxic.

The definitions are meant to protect majorities of populations but not necessarily other populations where circumstances are different or have changed. They are also not protecting the small percentage of individuals with different requirements due to genetic variations. Importantly, the definitions depend on our ability to determine effects with sound and meaningful measurements or observations. They are rarely linked to molecular functions.

The first and second regions in the dose/response curve, where deficiency and control of metal metabolism occur, are often considered in the realm of nutrition whereas the third region is in that of toxicology. Such a division is arbitrary because nutritional deficiencies affect the toxic response to other agents and an oversupply of one nutrient affects the status of others. A region of pharmacological action is rarely indicated. It should be related to a diseased organism. The meaning of "function" on the response axis usually remains unexplained. It often refers to sustaining life or causing death. All metal ions have multiple molecular functions. We usually do not know the hierarchy of these molecular functions and whether or not one molecular function is primarily responsible for the overall response. If function addresses a quantitative trait, such as growth, several molecular functions likely contribute to the overall effect. For example, compromised immunity is an early sign of zinc deficiency, indicating an essentiality for the immune system. However, functions in the immune system do not determine essentiality of zinc overall in the growth and maintenance of all cells. Cobalt in the form of vitamin B_{12}, on the other hand, participates only in two enzymatic reactions in humans and in this case it is more straightforward to relate its molecular functions to the overall effect of its deficiency. The toxicity associated with an excess of cobalt, however, is not readily explained based on single molecular targets. Thus, while there is specificity in using metal ions for biological functions, there is usually much less or no specificity in biological targets populated by an excess of metal ions and it is more difficult to pinpoint specific molecular targets of the toxic actions of metal ions. Only in some cases specific or preferential targets have been identified, e.g. δ-aminolevulinic acid dehydratase, a zinc enzyme involved in the synthesis of the porphyrin prosthetic group, as a target of lead toxicity. Because metal ions have multiple molecular functions, there is no reason why responses at high and low concentrations should be monophasic as given in dose/response curves. Multiple functions become compromised with acute or chronic exposure to an element or with its deficiency, both with graded severity for the health of an organism. A plateau of optimal functions may not reflect maximal functions, though. The latter may be reached only under ideal conditions and not in a normal environment because the organism needs to adjust to the availability of all essential metal ions and their interactions. An example is the activity of the selenium-containing enzyme glutathione peroxidase. Its activity increases upon supplementation with selenium but the increase does not mean that an organism is selenium-deficient when the enzyme has less than its maximum activity. Another characteristic of the plateau is that it can be rather narrow. The range between an element being deficient and being in excess can be very small. The behavior can be understood in terms of the necessity for tight control of metals and metal buffering, the basis of which will be discussed in detail in Chapter 6.

A dose/response curve depends on dynamic interactions of elements, the presence of other compounds, and multiple other factors such as disease or different types of stress. For instance, zinc deficiency is exacerbated by high amounts of phytate (inositol hexaphosphate) in the diet because phytate binds zinc and antagonizes its uptake. High availability of zinc in the gut, on the other side, antagonizes copper uptake and therefore may lead to copper deficiency. Anemia (iron deficiency) can occur due to a lack of copper or zinc because both are needed for iron metabolism: copper in enzymes catalyzing the oxidation of Fe^{2+} to Fe^{3+} and zinc in porphyrin biosynthesis. Non-essential elements also interact with essential ones. In anemia, cadmium uptake increases, making iron deficiency a risk factor for the toxicity of cadmium. Vice versa, the toxic effects of arsenic compounds can be lessened by zinc supplementation, which induces the synthesis of metallothionein that sequesters arsenic.

Despite the wealth of information we have about genes, their products and their functions, we do not seem to know with certainty all the elements that are essential for our health. We have limited knowledge about the functions of some elements, and consequently we do not know all the health implications of the essential elements or the non-essential elements which are also present. The previous chapter alluded to three major issues with presenting essential elements in a biological PSE. *First,* not all the elements are essential for one given species such as humans. "Essential" needs a reference to (a) specific organism(s). Some metals do not seem to be essential for all organisms (Ni, V) and yet have well established biochemical roles, and the essentiality of others seems to be restricted to few organisms in very special ecological niches with very specific functions (Cd, W). *Second,* our knowledge about the biological functions of several elements is still fragmentary. The discovery of an essential role of bromine in an enzyme involved in collagen metabolism in 2014 (McCall *et al.* 2014) is a stark reminder of the conclusion made over 30 years ago that the question which elements are important for life remains open-ended (Mertz 1981). In fact, molecular structures of nickel- or vanadium-containing enzymes in some species became available only after this statement was made. Controversies in accepting that some elements are essential, e.g. chromium, continue. *Third,* it needs to be understood what "essential" means. In the strictest sense, it means that the organism cannot survive when the metal ion is removed from the diet, i.e. the element is *nutritionally essential.* Even for some well-established essential elements such as manganese, such essentiality is difficult to demonstrate experimentally. The experiments are challenging, in particular with some ultratrace elements, where it is technically neither feasible to prepare synthetic diets that do not contain traces of these metal ions nor to control feeding of such diets without contamination with traces of these metal ions. There are also arguments in the literature that an element should be considered essential only if it has been shown to be *biochemically essential*, i.e. has an established molecular function. This argument is not factual and seems to reflect a perceived superiority of disciplines dealing with molecular structures over other disciplines dealing with function. Absence of evidence is not evidence of absence. The difficulties that had been encountered in revealing the complex molecular structure of the vitamin B_{12} cofactor seem to demonstrate that it is a mute argument. Demonstrating biochemical essentiality almost always lagged behind the demonstration of nutritional essentiality and functional effects. Establishing molecular structures linked to functions became possible only when the methods and methodology for investigations became available. Advances hinge on a better understanding of the definition of "essential". A broader definition had been suggested: "An element is essential when a *deficient intake* consistently results in an impairment of a function from optimal to suboptimal and when supplementation with *physiological levels* of this element, but not others, prevents or cures this impairment (Mertz 1981)". This definition gives the word "essential" a wider meaning beyond being essential for survival and includes functions that are a part of the system but nevertheless linked to significant health effects. The importance of fluoride in the form of fluorapatite, $Ca_{10}F_2(PO_4)_6$, for avoiding dental caries and supporting tooth health is an example. In this case, the term "essential" refers to a function that is much more narrowly defined and the term function is then clearly different from a nutritionally essential function.

The issue what "essential" means has come into focus again recently when it was suggested that some elements may be simply *beneficial*, improving some functions when supplemented (Nielsen 2014). A beneficial compound may have the same behavior in a dose/response curve as an essential compound but with the important difference that "function" on the response axis is a specific function not related to survival. In the older literature, one finds references to the potential essentiality of some other elements e.g. Li, As, Sn and even the ones that are considered primarily toxic, e.g. Pb, based on the observation that growth is affected when the diet is made deficient of the element. However, growth may be improved without the element being essential for growth. Metal ions have rather potent effects and can affect specific endocrine signals and signal transduction pathways involved in cellular proliferation. This issue of positive actions of primarily toxic metal ions will be resumed in the next chapter with a discussion of the concept of hormesis.

The wider definition of "essential" refers to *deficient intake*. It therefore becomes an issue how intake is determined. If it is measured, for example, by determining the metal ion concentration in blood plasma, the latter is dependent on the condition of the organism. During infections metal ions are removed from the blood plasma to avoid that they become available to invading organisms. Clearly, the removal does not reflect a lower intake. Increasing the intake under such a condition may be harmful. One can envisage other physiological responses where metal ions need to be lowered in a tissue to change a function during normal development of an organism. A correlation between function and *deficient intake* assumes that we know what the normal intake is in relation to optimal function. Clearly, metabolic needs change with conditions such as age, physical exercise, or disease. It requires a perfectly balanced diet to make all nutrients available at the same time. But even then it can be questioned whether our bodies always operate under optimal conditions for individual elements rather than striking a balance for all elements provided. Because of the interaction of elements, correlating function with variation of one element only can be misleading. Removing one element may result in a compensatory uptake of another element, so that the functional deficit is not due to the removal of the element in question but due to the effects caused by the uptake of another. Deficiencies have graded responses. The organism may give up some less important function for the sake of supporting more critical functions or may adjust in some other way to compensate for an emerging deficiency.

The wider definition also assumes that we know what *physiological levels* means in relation to *pharmacological levels*. Physiological level usually means a range in which the element is normally present in our tissues, but since concentrations can vary and since we do not know the full extent and functional implications of such variations, the border between physiological and pharmacological action can be blurred for some elements. Sometimes the terms beneficial and pharmacologic are used interchangeably.

Pharmacologic effects are usually at higher concentrations than the concentrations present in the diet or in tissues, but some metals have potent actions and a beneficial role that may be noticeable only once their concentrations are elevated only slightly. The distinction between beneficial and pharmacological is more difficult for elements that are already present at high concentrations but rather straightforward for elements that are virtually absent or present at extremely low concentrations such as platinum. Platinum compounds are anticancer drugs at therapeutic doses. Pharmacological activity of metal ions in the bottom half of the PSE can be more readily defined because of their special reactivity and because most of them are usually present only at very low concentrations.

This discussion may seem quite abstract but it illustrates the underlying assumptions and difficulties. One possible solution for the future is to relate dose/response curves to a molecular, biochemical concept such as the buffering capacity of a cell or a tissue for a particular metal ion and the range of its regulation (Chapter 6).

It is unknown to which extent a metal ion is essential for an organism that lives in symbiosis with another organism. A human being, for example, has more bacterial cells than own cells. Of course, bacterial cells are much smaller than our own cells. Nickel is essential for the pathogenic bacterium *Helicobacter pyloris*, but has no known molecular function in humans. Availability of nickel supports the growth of this ulcer causing/promoting bacterium and therefore nickel removal or restriction may be a specific means to starve this microorganism. Questions arise. Does this tenet hold for other elements and commensal bacteria? Are there elements that are only essential for bacteria that live in a host-guest relationship with humans or any other species? Can essentiality be indirect, namely by supporting the survival of an organism that is beneficial for the health of the host? We can expect some answers to these questions in the coming years.

The discussion centred on apparently healthy organisms. Genetic diseases where genes involved in metal metabolism are affected can cause metal deficiency or overload. In addition, changes of metal metabolism can be a consequence rather than a cause of the disease. In these cases, metal ions need to be given therapeutically, sometimes at pharmacological doses, or they need to be removed with chelating agents.

The matter of defining "essential" for species in terms of function becomes even more complicated because non-essential elements also can have biological functions and their presence can affect the function of essential elements as discussed in the following chapter.

4.2 The structures and functions of the essential metals

Short accounts of the essential and some non-essential metals will be given in this and the following chapter to provide an overview of their relative importance in biology. They are not comprehensive as detailed information can be found in textbooks of bioinorganic chemistry and monographs on the biology of individual metals. Metal ions have roles in enzymatic catalysis, including electron transfer and redox reactions, and in protein structure. In multicellular organisms, they have additional roles in generating electrochemical potentials at membranes and in cellular signalling.

4.2.1 s-block alkali and alkaline earth metals

These metal ions are often referred as "minerals" in nutrition and treated separately from the transition metal ions and zinc. From a standpoint of biochemistry, however, such a treatment is not well justified for magnesium and calcium because these divalent ions interact with other divalent metal ions, in particular zinc. In the order Na, K, Mg, Ca, the metal ions are increasingly important for controlling protein functions at specific metal-binding sites. The affinity for ligands is rather weak: for sodium (Na^+) and potassium (K^+) about millimolar, and for magnesium (Mg^{2+}) and calcium (Ca^{2+}) about micromolar, and hence the hydrated (aquo) ions are important in the solution chemistry. Though there is considerable similarity between sodium and potassium on one side and magnesium and calcium on the other, in biology very different roles have evolved. All four cations are tightly regulated.

Groups 1 and 2

Alkali group

Sodium and potassium. An important difference in the selection of the two elements is that sodium is extruded from cells whereas potassium has an inward gradient. Typical concentrations are 12 mM Na inside and 145 mM outside cells, while the situation is opposite for potassium: 140 mM K inside and 5 mM outside. These concentrations are regulated by hormones affecting the sodium pump (Na^+, K^+-ATPase) on the plasma membrane of cells. The concentration gradient is also important to control osmolarity and thus cell shape and to drive the transport of many substances through the plasma membrane. It is used to maintain electrochemical potentials across the plasma membrane of excitable cells, which for example is critical for the function of nerve cells (neurons), which are electrically excitable and transmit chemical information between neurons or between neurons and muscle cells (myocytes) through gaps, the synapse or the neuromuscular junction. In addition, both ions are used in the cell. Sodium has a role in regulating cellular calcium. Potassium is required for the activity of some enzymes.

Alkaline earth group

Magnesium and calcium. Magnesium is high inside (30 mM) compared to outside (1 mM) the cell whereas the total calcium concentration inside and outside are not too different (about 3 mM). The ions have quite different biological properties. As with Na and K, the ionic radii show an inversion: Mg < Ca for the un-hydrated ions, but Mg > Ca for the hydrated ions. Mg is kinetically more inert and often used as a cofactor or structural element. About 95% of Mg is bound to ATP. Mg also interacts with RNA. Mg and Ca are used as catalytic, structural and regulatory cofactors of proteins. Mg prefers octahedral complexes with a coordination number of six whereas Ca can adopt higher coordination numbers: Seven or even eight. Because these ions are divalent and have twice the charge of the alkali ions, they bind stronger to proteins and other ligands, and the distinction between bound and free becomes important. In the cell, free Mg^{2+} is 0.3 mM and free Ca^{2+} is 0.1 μM, i.e. free calcium is kept at much lower levels.

Calcium has an important additional function. It is a key cellular signalling ion that participates as a second messenger in information transfer and the control of some of the most important processes in metabolism, gene expression, cell motility, muscle and heart contraction, and bone mineralization. Intracellular Ca concentrations undergo spatially and temporally well controlled transients. Any perturbation of this control leading to calcium overload can result in disease and cell death. Calcium fluctuations in the cell are generated in response to hormones (first messenger) binding to the plasma membrane. The release of Ca in the cell is inositol-1,4,5-triphosphate (IP_3)/cADPR dependent and occurs from a store in the endoplasmic/sarcoplasmic reticulum (ER/SR). Ca^{2+} then binds to the protein calmodulin (CaM). CaM is a metal-activated molecular switch. It binds four calcium ions in EF hands, changes its conformation and then activates effector enzymes such protein kinases, adenylate cyclase, nitric oxide (NO) synthase, and Ca^{2+}-ATPase. Calcium release is also important for muscle function. In addition to the calcium traffic involving the ER, channels import calcium into the cell and export it from the cell via a membrane ATPase and Na, Ca antiport and into mitochondria via a Ca^{2+} importer. These proteins control calcium transients and keep the free calcium concentration very tightly controlled at about 100 nM. Regulation of calcium metabolism includes parathyroid hormone, a protein, and the hormone vitamin D, which increase absorption of calcium from the intestine and resorption from kidney and bone, where hydroxyapatite $[Ca_5(PO_4)_3OH]$ is the major mineral. As a consequence, calcium increases in blood. The protein calcitonin inhibits resorption and lowers calcium in blood.

4.2.2 d-block transition metals and zinc

With the exception of zinc, the metal ions are redox-active and thus catalysts for redox reactions over a wide range of redox potentials. In this period, the metal ions also have increasingly higher affinity for proteins. The metals at the beginning of the period form oxoanions in their highest oxidation states, thus they are anions not cations. Iron and zinc have by far the greatest number of functions. The use of copper and manganese is more restricted, and vanadium, molybdenum, cobalt and nickel all have rather specific functions in only a few enzymes. Functions of nickel and vanadium in human enzymes are not known. Chromium is a special case: It has been widely accepted as an essential element but such a role has been questioned recently. In contrast to all the other essential metals, a structure of a biological chromium complex is not known.

Groups 3 and 4

Metals in these groups are not essential.

Group 5

V: For the solution chemistry of vanadate in relation to phosphate, with which it competes in metabolism, it is important that its pK_a is about 1.5 units higher than that of dihydrogenphosphate ($H_2PO_4^-$), i.e. $H_2VO_4^- \rightarrow HVO_4^{2-} + H^+$, $pK_a = 8.2$. V(V) in vanadate can be reduced to V(IV) in the form of oxovanadium(IV) ("vanadyl" cation, VO^{2+}) which binds to transferrin, and under some special circumstances reduced further to V(III). Vanadium(V) is a cofactor in haloperoxidases in some bacteria and fungi. These enzymes use vanadate in the active site bound to an imidazole nitrogen of histidine and halogenated organic substrates with hydrogen peroxide forming a hydroperoxido intermediate and a halide ion (chloride, bromide, iodide). The redox state of vanadium does not change in these reactions. A vanadium-containing nitrogenase exists in bacteria and in this case the vanadium replaces the molybdenum usually found in the cofactor of these enzymes. In humans there is only about 1 mg vanadium in the body of an adult and there is no known molecular function that would indicate that vanadium is essential. Due to the similarity to the phosphate ion, vanadate and vanadium compounds inhibit phosphatases, a property that is most pronounced for vanadate but shared with other oxoanions. Vanadium compounds stimulate the insulin response, hence are insulin-mimetic or insulin-sparing, also through a mechanism involving competition with phosphate for enzyme active sites and inhibition of their actions. A specific vanadium(IV) complex (amavadin) is found in the fly agaric (*Amanita muscaria*). It has two 2,2'-(oxoimino)dipropionate ligands and is extremely stable. Vanadium is also present in specialized cells of sea squirts (tunicates) as $[V(III)(H_2O)_5HSO_4]^{2+}$. The function of vanadium in these species is not known.

Group 6

It is the only group where metals from all three periods are being used in organisms.

Cr: There are two stable oxidation states and they have completely different chemical and biological properties. Chromate (CrO_4^{2-}), Cr(VI), is an anion, relatively toxic, and a carcinogen. Cr^{3+}, however, is a cation and thought to be essential for glucose tolerance. However, the notion that it is an essential nutrient for humans has recently been challenged (Vincent 2013). Compared to all other essential metal ions, chromium(III) complexes are kinetically rather inert. They exchange their ligands extremely slowly. Accordingly, the biological chemistry of chromium is linked to its complexes.

In the early 1950s, diets were investigated with regard to whether or not they provide all the necessary nutrients. Male Sprague-Dawley rats developed progressive glucose intolerance, liver necrosis and finally death when they were fed a diet that consisted of 30% *Torula* yeast. The diet was low in sulfur amino acids, vitamin E, and factor 3, which was later identified as selenium. However, glucose tolerance remained normal or even was restored when a small amount of brewer's yeast or pork kidney powder was added to the deficient diet. A factor termed glucose tolerance factor (GTF) was separated from factor 3 and turned out to be heat stable after ashing, suggesting an inorganic compound. It contained chromium(III). Several chromium compounds then were found to be active in this rodent assay for glucose tolerance. Leading work in animal and human nutrition for the next 40 years resulted in accepting chromium(III) as an essential nutrient. However, it was made clear that three experimental challenges remain, namely (i) the structure of the biologically active chromium complex, (ii) ways to determine chromium status in humans, and (iii) to determine chromium's exact mechanism in insulin action on glucose and lipid metabolism (Mertz 1993). Part of the controversy that ensued about whether or not chromium should be continued to be considered as an essential element arose from the frustration about not being able to meet these challenges and answer rigorously the scientific issues at hand. There seems to be no evidence for chromium deficiency and beneficial effects of chromium supplementation in healthy populations but there is evidence on improvement of function when supplementing some but not all diabetic individuals with chromium(III) compounds. The concentrations of chromium are as low as those of molybdenum and cobalt, both of which are established essential trace elements. The absence of a structure and a mechanism to explain its action do not prove that it is not an essential element.

Sixty years of biological chromium research and a vast literature demonstrate the experimental difficulties in determining the function of chromium(III) and highlight the issues with the meaning of "essential."

Mo: There are only a relatively small number of molybdoenzymes. In humans there are only four: sulfite oxidase, xanthine oxidase, aldehyde oxidase, and the mitochondrial amidoxime reducing component (mARC) belonging to three protein families. However, in other organisms there are additional molybdoenzymes belonging to another two protein families, one of which contains tungsten instead of molybdenum as cofactor. In all these families, molybdenum is bound to a pterin cofactor (Moco) with an ene-1,2-dithiolate (dithiolene) functional group (Figure 2.5B). The only exception is nitrogenase with its cofactor (FeMoco). In this cofactor, which has the composition $MoFe_7S_9C$, Mo is bound in one position where an iron normally would reside in the two 4Fe–4S clusters and is also bound to homocitrate. A characteristic feature of the cluster is a carbon atom (Figure 2.7B).

Similar to diseases arising from mutations in enzymes of the porphyrin biosynthetic pathway, there are diseases that lead to molybdenum cofactor deficiency. The initial metabolite for the synthesis of the cofactor in at least six steps is believed to be GTP. Cells take up molybdenum in the form of molybdate (MoO_4^{2-}) using a specific transporter.

W: Tungsten replaces molybdenum in enzymes in special organisms (thermophilic archaea), e.g. formate dehydrogenase. The essentiality of W is restricted to only specific biological niches because the availability of the two elements has a different pH dependence. Tetrathiotungstate is soluble whereas tetrathiomolybdate is insoluble at neutral pH. In light of the otherwise very similar physicochemical properties of molybdate and tungstate, it is most remarkable that organisms have developed mechanisms of selecting either one or the other.

Group 7

Mn: The number of manganese proteins in a given organism remains unknown. In part, this is due to a lack of consensus sequence motifs for manganese binding sites, precluding predictions by bioinformatics approaches. Concentrations of manganese in humans are rather low, but manganese is believed to be a cofactor in many enzymes. Mitochondrial superoxide dismutase (SOD), arginase, glutamine synthase, pyruvate carboxylase, and lysosomal enzymes are examples. The pro-antioxidant effects of manganese are thought to be related to the fact that it does not perform the redox chemistry typical for iron. Otherwise the sizes of the Mn^{2+} and Fe^{2+} ions and their affinities for ligands are rather similar, with the notable exception that iron has a tendency for coordination with sulfur ligands whereas manganese prefers oxygen ligands. In humans, no specific transporters for Mn^{2+} have been identified. However, the ion uses iron import systems (transferrin and DMT-1), iron exporter (ferroportin), zinc importers (Zip) and exporters (ZnT), and calcium importers. Given the selectivity of membrane transporters (Chapter 6) and the need for regulating manganese concentrations, such a role of transporters for other metal ions in manganese transport and the apparent lack of specificity seem unexpected. Maybe, the transport properties do not reflect the situation *in vivo* and we have not yet identified the putative manganese transporters, or there are yet other unknown mechanisms for controlling manganese. If exposure to manganese compounds leads to an overload, the consequences are manganism, a condition with symptoms similar to Parkinson's disease with degeneration of dopaminergic neurons.

A remarkable manganese cluster is found in the water-oxidizing enzyme of photosystem II of plants (OEC = oxygen evolving complex) (Figure 2.7). It contains four manganese ions and one calcium ion. In the reaction in which molecular oxygen is made from water, a very strong oxidant needs to be generated: It is in the form of Mn in higher oxidation states of Mn(III) and Mn(IV).

Group 8

Fe: Next to zinc, iron participates in the largest number of metalloproteins. Iron is taken up as either heme or the free ion. In the non-heme dependent pathway, Fe(III) is reduced to Fe(II) at the apical surface of intestinal cells (enterocytes) by the reductase Dcytb and taken up by the transporter DMT-1 (divalent metal transporter-1). It is then stored in ferritin, used in the cell, or exported into the blood at the basolateral site of enterocytes by ferroportin. Outside the cell it is oxidized by the copper-dependent ferroxidase hephaestin. In blood, Fe(III) is bound to transferrin with very high affinity. Cells take up the iron-loaded transferrin through transferrin receptor-mediated endocytosis, Fe(III) is released in the endosome, reduced to Fe(II), and transported by DMT-1 into the cell. Aside from storage in ferritin, it is channelled into at least three different types of proteins: heme proteins, Fe–S proteins and binuclear and mononuclear non-heme iron proteins, for which predictions suggest about 300 human proteins. Heme proteins make a large class of different types of proteins, e.g. cytochromes, hemoglobin, and myoglobin. Heme biosynthesis occurs in mitochondria. Iron is inserted into protoporphyrin IX by the enzyme ferrochelatase to form heme. Fe–S cluster assembly occurs in mitochondria and in the cytosol and involves several specific proteins. Storage of iron is unlike that of any other metal ion. It occurs in ferritin, which oxidizes Fe(II) to Fe(III) at a ferroxidase site and deposits the iron as iron oxide in a core that can contains >3000 iron atoms. Reduction of ferric to ferrous ions can mobilize iron from ferritin through a process that is not well understood. Thus, the change of valence state between ferrous and ferric determines its availability and mobility: Fe(II) is mobile while Fe(III) is immobile. In enzymes, Fe(IV) and Fe(V) oxidation states can occur transiently to generate powerful oxidants.

Free iron needs to be tightly controlled as it participates in non-enzymatic redox cycles that generate damaging hydroxyl radicals:

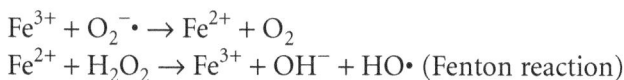

$$Fe^{3+} + O_2^- \bullet \rightarrow Fe^{2+} + O_2$$
$$Fe^{2+} + H_2O_2 \rightarrow Fe^{3+} + OH^- + HO\bullet \text{ (Fenton reaction)}$$

The sum of these two reactions is the Haber–Weiss reaction, the iron-catalyzed formation of the hydroxyl radical:

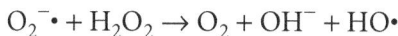

$$O_2^- \bullet + H_2O_2 \rightarrow O_2 + OH^- + HO\bullet$$

Numerous diseases are associated with iron deficiency or iron overload. Primary iron overload is due to hereditary disease (hemochromatosis) whereas secondary iron overload describes other conditions that increase the body's iron burden. Paradoxically, they can lead to iron loading anemias different from iron deficiency anemias.

Group 9

Co: Cobalt is essential as a cofactor in the form of vitamin B_{12}. Cobalamins are tetrapyrrole derivatives (corrins). Vitamin B_{12} is used in only two enzyme reactions in humans, methylmalonyl-CoA mutase, converting methylmalonyl-CoA into succinyl-CoA by using adoB12 and methionine synthase, converting homocysteine into methionine by using MeB12. The sixth ligand gives the name, ado = adenosylcobalamin, i.e. the ligand is 5′-deoxyadenosyl) and Me = methylcobalamin. The vitamin B_{12} cofactor is utilized in additional enzymes in bacteria, which have the ability to synthesize vitamin B_{12} in aerobic and anaerobic pathways requiring up to 30 enzymes. Insertion of cobalt into cobalamin requires energy in the form of GTP. Deficiency of vitamin B_{12} leads to pernicious anemia. The reason for developing anemia is the role vitamin B_{12} in methionine synthase, which is part of folate metabolism. Vitamin B_{12} deficiency limits the folate required for the development of erythrocytes, which look larger than normal, hence the name megaloblastic anemia for the condition. The effect of vitamin B_{12} deficiency is not a direct effect on iron metabolism but rather on a cell type needed in iron metabolism, providing a remarkable example of how an indirect interaction of elements can have serious effects. Cobalt is redox active and three valence states are employed in biology: Co(I), Co(II), and Co(III). In humans, there are no known additional reactions requiring non-corrin cobalt. In fact, additional cobalt(II) in non-corrin form is quite toxic due to its redox activity. However, less than a dozen of non-corrin cobalt-requiring enzymes have been described for bacteria. Nitrile hydratase is a Co(III) enzymes. Co(III) is rather inert kinetically (eight orders of magnitude slower ligand exchange than iron(III)), yet active, but it is unclear whether it is a case of metal switching where Co(III) substitutes for Fe(III) in this enzyme.

Group 10

Ni: Nickel has no known function in human proteins. The first nickel enzyme identified was urease from jack beans. Nickel forms a binuclear site in this enzyme. An additional eight nickel enzymes were then identified in bacteria and archaea: methyl coenzyme M (2-mercaptoethane sulfonate) reductase, carbon monoxide dehydrogenase, SOD, acireductone dioxygenase, glyoxalase I, acetyl CoA synthase, lactate racemase, and hydrogenase. Nickel insertion needs metallochaperones and nickel acquisition apparently employs nickelophores. Nickel is also part of a tetrapyrrole-based cofactor (hydrocorphin, F430, Figure 2.5A).

Group 11

Cu: Copper is a cofactor in a number of redox enzymes. The uptake transporter is Ctr1 (copper transporter 1). Cu(II) is reduced to Cu(I), which is the form transported and the major valence state of copper in the reducing environment of the cell. Cu(I) is handled by specific metallochaperones to avoid redox chemistry of the highly reactive free ion and to supply copper to Cu, Zn SOD, cytochrome c oxidase in mitochondria, and two copper exporters, ATP7B in liver and ATP7A in the intestine. Genetic diseases associated with copper metabolism have mutations in these exporters: Wilson's disease, which results in copper overload due to mutated ATP7B and decreased secretion of copper into the bile, and Menkes disease, which results in copper deficiency due to mutated ATP7A and diminished supply of copper to tissues and copper-requiring enzymes. Three types of copper centers were identified based on their spectroscopic characteristics: type 1, type 2, and type 3. In cytochrome c oxidase, the major oxygen reducing site contains a binuclear copper where the valence state of each copper is formally 1.5.

Group 12

Zn: The zinc concentrations in cells are remarkably high, about 200 μM, putting them in the order of the concentrations of major metabolites such as ATP and emphasizing that at the cellular level we are not dealing with a trace metal. Also, like iron, the total amount in a human is a few grams.

Zinc is colorless and strictly speaking not a transition metal ion. The lack of physical properties that are typical for the transition metal ions and allowed their characterization by spectroscopic techniques made zinc a less favoured metal ion among bioinorganic chemists. Accordingly, zinc biochemistry developed comparatively late because there were fewer methods to investigate cellular zinc redistribution. Only with the advent of synthetic chelating agents that fluoresce upon zinc binding a powerful tool became available to investigate its redistribution. In contrast to its lack of physical properties typical for other transition metal ions, among all the metals in biology, its chemical properties make it the metal ion 'par excellence' employed for the largest number of biological functions. It is a Lewis acid (an acceptor of electrons), stereochemically has coordination flexibility, and like Mg and Ca it is redox-inert. In contrast to calcium, which does not coordinate with sulfur ligands in biology, sulfur (cysteine) coordination environments in zinc sites allow for ligand-centered oxidoreduction with concomitant zinc binding and release. This particular chemistry links zinc and redox metabolism despite the fact that zinc is redox-inert. It makes zinc mobile from sites where it is tightly bound. Moreover, oxidative release of zinc during redox signalling generates zinc(II) ions that bind with rather high affinity to proteins. This mechanism endows zinc with effector roles.

Hardly any cellular process does not depend on zinc. It is widely used in all classes of enzymes as a catalytic, structural, or regulatory metal ion, particularly in metalloproteinases. As a structural component of proteins, it makes an enormous contribution to the variations in protein structures as it organizes protein domains to yield structures that cannot be achieved in its absence. The zinc-binding domains are important for protein–protein interactions or protein–DNA/RNA interactions. Many of the structural sites with sulfur coordination are collectively referred to as zinc finger proteins (Figure 2.15).

In addition to catalytic and structural roles, a third area is its role in regulation. Like calcium, it is a signalling ion and also released from an ER store. Phosphorylation of the transporter/channel Zip7 is involved in this process. It has different and complementary roles to calcium. It binds stronger to targets and selects different targets. Signalling also involves release of zinc from vesicles to the extracellular space. Examples include vesicular exocytosis of zinc in the brain, where it is a synaptic transmitter in neurons with zinc-rich terminals (boutons), particularly in the hippocampus; release into the mother's milk; and release from the granules of pancreatic β-cells where it is stored in a complex with insulin. Thus, it has paracrine (affecting another cell nearby) and autocrine (affecting the same cell), and maybe even endocrine roles (like hormones affecting cells far away).

4.2.3 p-block metals

None of the metals are essential.

4.2.4 f-block metals

Very recent research suggests that cerium, a lanthanide element, serves as a cofactor in the methanol dehydrogenases of methanotrophic life in some volcanic mud pots (Pol *et al.* 2014).

Summary

The chapter started with a discussion of the dose/response curves for essential metal ions, emphasizing the meaning of the functional response and the fact that doses and responses are species-specific. It continued with providing a working knowledge of the role of metals ions in biology. For a wider acceptance of the important roles of metal ions in biochemistry, it must be understood which functions are general to most or all forms of life and which ones pertain to only a few special organisms. At least 12 metal ions are essential for life. The exact number of essential metals for humans and most other organisms is not known as there are controversies and uncertainties. The essential metals include the alkali metal ions sodium and potassium and the alkaline earth metals magnesium and calcium, the transition metal ions molybdenum, manganese, iron, cobalt and copper, and zinc. Nickel and vanadium are essential for only some organisms and controversy surrounds the status of chromium as an essential trace metal. Additional metal ions are utilized by very specific organisms. With the exception of chromium, molecular functions in proteins are known. The functions of essential metals range from just a few enzymatic reactions for some metals (cobalt and molybdenum) to thousands of functions for others (zinc and iron). The use of metabolic energy (GTP or

ATP) in metal metabolism and the use of metal ions in cellular regulation demonstrate their integration in general metabolism. It is unlikely that the additional metal ions that are present are functionally inert: Each metal ion has some biological effect, some are beneficial and not necessarily essential, and all become toxic at high concentrations. Metal ions interact and the status of one essential metal ion affects that of others as well as the biological actions of non-essential metal ions. Relating macroscopic observations of metal deficiencies and overload to molecular functions is an ongoing effort and an area where a systems approach such as metal-lomics is called for. Relating the structure and function of isolated biomol-ecules to the biology of the entire organism and vice versa needs to consider interactions among metal ions, how biological control limits metal concentrations, and the additional biological functions that metal ions gain in biological systems.

General references

A. Sigel, H. Sigel, R.K.O. Sigel, eds., Interrelations between Essential Metal Ions and Human Diseases. Metal Ions in Life Sciences, 13, Springer 2013.

L. Rink, Zinc in Human Health and Disease, IOS Press 2012.

R. Crichton, Iron Metabolism: From Molecular Mechanisms to Clinical Consequences. 3rd edition, Wiley 2009.

Specific references

A. Pol, T.R.M. Barends, A. Dietl, A.F. Khadem, J. Eygensteyn, M.S.M. Jetten, and H.J.M. Op den Camp (2014). Rare earth metals are essential for methanotrophic life in volcanic mudpots. *Env. Microbiol.* 16, 255–264.

A.S. McCall, C.F. Cummings, G. Bhave, R. Vanacore, A. Page-McCaw, and B.G. Hudson (2014). Bromine is an essential trace element for assembly of collagen IV scaffolds in tissue development and architecture. *Cell* 157, 1380–1392.

F.H. Nielsen (2014). Should bioactive trace elements not recognized as essential, but with beneficial health effects, have intake recommendations. *J. Trace Elem. Med. Biol.* 28, 406–408.

J.B. Vincent, The Bioinorganic Chemistry of Chromium, Wiley 2013.

W. Mertz (1981). The essential trace elements. *Science* 213, 1332–1338.

W. Mertz (1993). Chromium in human nutrition. *J. Nutr.* 123, 626–633.

Chapter 5

Non-essential elements

The previous chapter discussed uncertainties in considering some elements essential and how many elements are essential for a given species. The "scientific jury is still out" whether additional metals are essential and what the functional implications of their presence are. The history of scientific discovery teaches us that one ought to be careful when assuming that an element is non-essential. About 90 years ago, when the biological functions of zinc in humans were not yet known, it was discussed whether or not the presence of zinc in biological tissues is a consequence of industrial exposure to zinc. A landmark paper concluded that the omnipresence of zinc in human tissues indicates a critical function for the body (Drinker and Collier 1926). It took another 20 years to isolate the first zinc enzyme, carbonic anhydrase, and 40 years to demonstrate that zinc is an essential nutrient for humans.

Clearly, not all chemical elements are essential for life. This chapter will continue the discussion by focusing on the remainder of the elements considered to be non-essential. Lumping together elements that are not proven to be essential in the category "non-essential" should not imply that these elements do not have biological functions. In addition to the essential metals, many other metals are always present in humans but we do not know their functions. Notably, titanium, rubidium, strontium, aluminium, cerium, tin, barium and the toxic metals lead, and cadmium are all present at significantly higher concentrations than those of the essential metal ions molybdenum and cobalt or the essential elements iodine and selenium. For some non-essential elements, there are remarkable bioaccumulation factors. Concentrations of others can be rather constant and reflect the occurrence in the environment, e.g. lanthanides, or they increase during lifetime, e.g. cadmium, which accumulates with yet to be fully determined consequences for our wellbeing and the process of ageing. We need to know at which concentrations elements are normally present, whether or not they accumulate, what their metabolism is, at which thresholds they affect biological function, and what their molecular targets are and how these targets relate to overall effects. If a metal is not taken up for a function, the uptake likely reflects a lack in selectivity of the uptake systems of essential elements and piggy-backing on these systems. Metal cations with similar size and charge and metal oxoanions that resemble phosphate or sulfate show such molecular mimicry.

A major aspect of many metal ions in comparison to other compounds is their exquisite chemical reactivity due to redox and catalytic activity and high affinity binding to proteins. They are therefore bioreactive elements and not innocent bystanders that are present without any functional outcome (Cvetkovic *et al.* 2010). *The concept of bioreactivity is important because it addresses the biological functions of all the elements in the periodic table, in particular those that are not generally discussed as part of our biochemistry, thus extending the biochemistry of metals beyond that of the essential ones and including many more elements.*

Biological responses to metal ions will vary among species, not only in a quantitative way due to different levels of adaptations but also in a qualitative way due to different usage of metals. Responses vary depending on the nutritional and health status, and genetic differences. In the experimental sciences, effects are usually investigated over a relatively short period of time, i.e. acute effects. Long-term, chronic effects over years or decades of human life are rarely investigated for obvious reasons. Metal ions are not biodegradable and may accumulate over long periods of time. There is sufficient evidence that some metal ions — or imbalances due to deficiencies and overloads of essential metals — generate redox stress with ensuing DNA damage with long term, negative impact. The limited knowledge in this area and the lack of biomarkers and clinical tests for determining the nutritional status of many metals is a serious issue for human health. For the non-essential elements there is an even greater scarcity of data than for the essential elements, and in some cases, an almost complete lack of information regarding the long-term health effects. Furthermore, our manufacturing practices have harnessed the useful properties of elements in the lower half of the PSE and we are now increasingly exposed to some metals and chemical species such as fabricated nanoparticles, which have not been part of our previous environment and for which adaptations have not evolved. We have had limited opportunities to evaluate biological effects.

5.1 Dose-response curves for non-essential metal ions

The dose-response curve for non-essential elements is different from that of essential elements as there are no signs of deficiency when concentrations are lowered (Figure 5.1). The behavior in terms of life and death of an organism is usually described with a region of "no effect" followed by toxicity at higher concentrations. However, if functions other than life and death are considered, the situation is more complex and a description with a curve containing only two regions is inadequate. There can be positive effects if the metal has some beneficial or pharmacological effect before negative effects set in and toxicity occurs at higher concentrations.

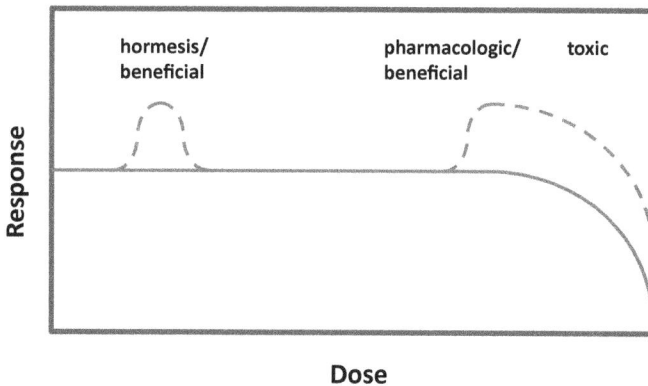

Figure 5.1   Dose/response curve for non-essential metal ions. The idealized curve is usually given with two regions only: a no-effect region where the organism tolerates the metal, followed by toxicity. However, there are two additional effects: hormesis at low concentrations, where a metal ion may have a stimulatory effect, and a pharmacological effect at higher concentrations before the metal becomes toxic.

Thus, at higher concentrations metals need to be discussed in terms of pharmacology (in a diseased state) and toxicology (in a diseased or in a healthy state), respectively.

A "no effect" relationship before a threshold is reached and toxic or beneficial effects occur may be observable at the level of an organism depending on what is being measured but at the molecular level it is difficult to envision that a metal ion is inert and does not interact with biomolecules without functional consequences. Some metals may simply mimic a function of another element present with similar chemistry (e.g. Rb for K, Sr for Ca). The linear region can be small or large if there is some form of tolerance or mechanism of detoxification (Chapter 6) where the organism can cope with the metal ion until a certain threshold is reached. Above the threshold higher concentrations improve or compromise function. Usually, dose-response curves describe the response of *added* metal ions. However, most metal ions are already present at some concentration, and one also needs to consider the functional consequences of removing the endogenous metal with a chelating agent (chelation therapy) or of lowering its concentration by changing the diet or the exposure. In an organism that has already been exposed to a primarily toxic metal, *removal* of the metal may have a beneficial effect. In contrast, removal of an essential element will have a detrimental effect. However, if a non-essential metal ion has a beneficial function, function may also be compromised when intakes are lowered or the metal removed by other means. In this case, the dose/response curve will resemble that of an essential element (Chapter 4).

Dose-response curves may not be linear (monophasic) and nonlinearity is an issue when evaluating the risk of exposure. It is often assumed that the lower the concentrations the lower the risk. The non-linearity with a positive effect before a negative effect sets in is referred to as *hormesis*, a phenomenon widely described for many substances but not accepted by some in the scientific community (Calabrese 2013). The biphasic effect is due to the fact that biological effects can be additive (synergistic) or compensatory (antagonistic) at different concentrations. Responses are based on observations of entire organisms but they are multifactorial as already discussed for the pleiotropic molecular actions of essential metals. The overall response in the organism is the sum of all molecular actions, many of which we may not even understand.

In the dose-response curve at higher concentrations, there are three actions, two are positive (stimulatory), and one negative. Eventually all metal ions including the essential ones become toxic at high concentrations.

5.1.1 Bioactive metal ions — beneficial effects

The distinction between beneficial and essential is a matter of whether or not the function addressed is one critical for survival. Deficiency of an essential element eventually results in death whereas the deficiency of a beneficial element merely compromises a function (Chapter 4). A beneficial effect occurs when a given amount of the element in the diet optimizes a particular function, or functions, in a healthy person. Health benefits may include protection against the risk of developing disease. The definition has been discussed for Cr, Ni, B, and Si, where beneficial effects have been demonstrated in reducing the risk for chronic disease, such as osteoporosis (Si) and diabetes (Cr), and in supporting development of the nervous system and bones (B) and various other positive effects (Ni) (Nielsen 2014). In the case of these elements, their removal from the diet has not been shown to result in death and molecular functions in biomolecules that would explain the beneficial functions are not known *for humans*. Bioactivity is primarily thought to be an action at nutritional levels, suggesting that recommended dietary intakes should be made and the status monitored (Nielsen 2014). The distinction from essential elements may be apparent only as it attests to a lack of knowledge that would allow us to classify these elements as either essential or pharmacologically active.

5.1.2 Pharmacologically active metal ions — beneficial effects

Pharmacological effects could be mistaken for beneficial effects as the distinction hinges upon what defines a healthy organism and a normal function. A beneficial effect improves or optimizes a function in an otherwise healthy organism whereas a pharmacological effect is seen when a disease is treated and an organism restored to normal function. Side effects of most therapeutic drugs illustrate the principle that multiple molecular targets exist and that improvement of one function can be accompanied by compromising others. A pharmacological action usually occurs before an element becomes toxic. The famous adage of Paracelsus (Philip von Hohenheim, 1493–1541) makes this point: *it is the dosage that makes a substance either a poison or a remedy*. Thus, one can employ a toxic effect pharmacologically, e.g. in platinum drugs as anticancer agents.

Essential elements are used as pharmaceuticals as well. They may have pharmacological actions that are not considered part of their action spectrum. Metals are pharmaceuticals when they are used to treat a metal (nutrient) deficiency. Nutrients that have therapeutic actions are called *nutraceuticals*.

5.1.3 Toxicity

Quoting Paracelsus again is appropriate as a way of introduction: "Poison is in everything and nothing is without poison". While a pharmaceutical response is usually due to selected targets, the toxic response affects many systems as a function of the dose and hence it is not necessarily monophasic either. A term often used to describe toxicity is the LD_{50} value, the lethal dose that results in the death of 50% of the organisms. The value refers to acute toxicity. Chronic toxicity is more difficult to fathom, though, and causation more difficult to prove. It includes mutagenicity and ensuing carcinogenesis and the development of other chronic diseases.

The discussion dealt primarily with metal ions prevalent in the biological milieu. Chemical species of metal ions in other oxidation states and different coordination environments have different toxicological profiles. This applies in particular to elemental metals and organometallic compounds. They have different reactivity and modes of uptake, and if dispersed, e.g. inhaled in dust, taken up in the intestine as small particles or in contact with the skin, engender completely different toxicology. The toxic response is influenced by additional factors such as the nutritional status of essential elements, demonstrating the interdependence of essential and non-essential metal ions. Deficiency of an essential metal can be a cumulative risk factor for the action of a toxic metal. For example, zinc deficiency causes oxidative stress, which can be augmented through exposure to another metal that causes oxidative stress. Likewise the uptake of cadmium increases under conditions of iron deficiency, making iron deficiency a cumulative risk factor for cadmium toxicity. Toxicity of metal ions depends on the pathways of uptake and distribution of metal ions and can be exacerbated if the selectivity governing uptake is by-passed. One example is intravenous injections where bio-availability of essential and other metal ions is much higher than in the diet and where contamination with metal ions can occur. In the intestine and in the lung, barriers control which metal ions will end up in the blood; and there are additional barriers with selective permeabilities protecting additional organs, e.g. the blood brain barrier, the placental barrier and the blood testes barrier in reproductive organs, and the blood retinal barrier in the eye. The liver has a specific role in intoxication and detoxification by handling metal ions that reach it from the blood. Some metal ions are excreted into the bile. The kidney also has a role in detoxification but it is also particularly vulnerable to metal toxicity.

The toxic response varies between organisms. An element that is mostly poisonous for most organisms can be employed for a biological function in another. An example is the use of cadmium instead of zinc in carbonic anhydrase in some marine organisms. There are windows of susceptibility during development where the toxicity is much higher, thus age is a factor in how toxicity is expressed.

Toxicity depends on whether or not the organism has developed specific mechanisms of detoxification (Chapter 6). In contrast to essential metal ions which are specifically allocated to their targets in the range where regulation of metal metabolism takes place, toxic metal ions often show an apparent lack of specificity. This appears to be the case for cadmium which targets a plethora of proteins. In the case of other metal ions more specific pathways for the toxic action have been identified. They involve binding to specific biomolecules such as proteins or DNA and altering signal transduction and gene expression, generating oxidative stress or leading directly to DNA damage. Often, a toxic metal replaces an essential metal and inhibits its function. Lead, for example, targets specifically 5-aminolevulinic acid dehydratase and binds in the catalytic site instead of zinc. For many other toxic metals, however, the overall signs of toxicity have not been correlated with individual molecular targets. Metal ions that are "foreign" to the organisms can interfere with the action of essential metal ions, for instance by having higher affinities to target proteins, or they bind to proteins that are not metalloproteins. In the latter case, there is *an additional bioinorganic chemistry of metal coordination in proteins beyond the classic metalloproteins. Many enzymes employ histidine, glutamate or cysteine side chains in their active sites for catalysis. Catalysis requires proximity of these residues, which therefore provide potential metal binding sites in these enzymes.*

Metals can become available at variable concentrations through exposure, which is either inadvertent (poisoning) or specific (environmental, therapeutic, or industrial). The discussion is not restricted to the non-essential elements because essential elements also become toxic at higher concentrations, and high availability of one can condition the deficiency of another. While there is control of essential elements, we are unprotected if this control is overwhelmed, and we are even less protected for those non-essential metals for which there is no control. We were always exposed to metals, even to those from the toxic triad (Cd, Hg, Pb) through the soil and the water from which our food originates as well as depending on how our food is processed and prepared, and also from the air we breathe. There can be considerable accumulation of metals in plants and animals and there is additional accumulation through food chains and food webs. The exposure has changed over time with industrialization and is changing again now with new types of materials and procedures employed in manufacture. There have been cases of serious industrial contaminations with metals. Methyl mercury (Minamata disease) and cadmium (Itai-Itai disease) poisoning of populations may serve as a reminder of the dangers. They are by no means the only examples. There are many more cases, some on a more global scale, where the concentrations of metals were too high, e.g. lead, where crops growing on contaminated soil or seafood raised in contaminated waters became poisonous, and where metal contamination of drinking water exceeds the legal limits. Industrial hygiene has improved enormously but the accumulation of metals in the environment continues to be a major concern, in water, in the soil, but also in the air. The air quality in terms of metal particles in some of our most densely populated cities is a significant threat to our health. For some long known metal toxicants, i.e. substances we have been exposed to for eons, some mechanisms of detoxification exist. However, more recently, we are becoming exposed to additional elements for which no protection has evolved. There are new chemical species, such as nanoparticles and nanomaterials with yet other routes of uptake and unique chemical reactivities. Not only are there uncertainties as to whether we have indeed identified all the essential elements, but we know even less about the biological effects of the remaining elements with many of which our bodies have to deal now.

5.2 Metals in medicine

Some metal ion complexes are potent drugs because of their reactivity and binding affinity, and because the metals normally are not present at concentrations where the pharmacological activity is expressed. In addition to the medicinal use of some essential and non-essential elements in the top half of the PSE, mostly the heavier elements in the bottom half feature prominently in the pharmacopeia of metallodrugs. The discipline has been called medicinal inorganic chemistry as it offers some advantages to medicinal organic chemistry, the traditional playground of drug discovery (Mjos and Orvig, 2014). A medical periodic table has been constructed to illustrate the use of many metal ion complexes as metallodrugs in therapy to treat infections, inflammation, and cancer and in diagnosis (Barry and Sadler 2013; Chellan and Sadler 2015).

5.2.1 Therapy

Metallodrugs. A few now classic examples are As, Pt, Bi, and Au drugs. Paul Ehrlich (1854–1915) introduced an arsenic compound based on 3-amino-4-hydroxyphenyl arsenic(III) (tradename Salvarsan) to treat syphilis (Figure 5.2).

It is thought to be a mixture of trimeric and pentameric species. Arsenic trioxide is used to treat acute promyelocytic leukemia. Barnett Rosenberg (1926–2009) discovered a potent anti-cancer metallodrug: cis-platin, from which carboplatin (Figure 5.2) and later generations of platinum drugs were derived. Cis-platin and related compounds have exchangeable ligands in cis. Purine bases of DNA can bind to these coordination sites leading to DNA intrastrand crosslinks that cause cell cycle arrest and apoptosis. Gold(I) compounds such as aurothiomalate, aurothioglucose and auranofin (Figure 5.2) are anti-inflammatory and used to

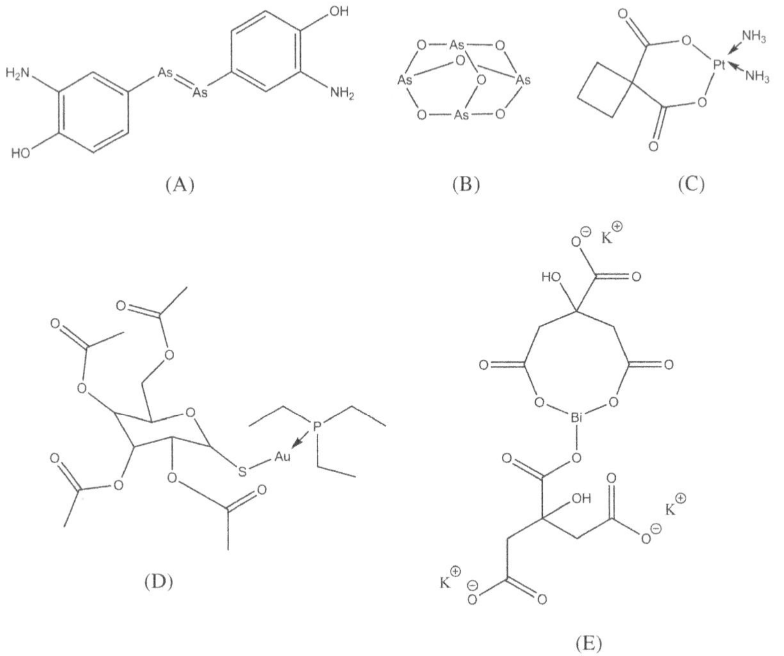

Figure 5.2   Examples of metallodrugs. (A) As (a metalloid): Arsphenamine, tradename Salvarsan, was used for the treatment of syphilis and trypanosomiasis. It is a mixture of a trimer and a pentamer (not shown); (B) As: Arsenic trioxide (As_2O_3), tradename Trisenox, is a legendary poison but used for treating certain leukemias; (C) Pt: Carboplatin, tradename Paraplatin, is a derivative of cis-platin and used in chemotherapy to treat many types of cancer; (D) Au: Auranofin, tradename Ridaura, is used to treat rheumatoid arthritis; (E) Bi: Bismuth subcitrate, tradename De-Nol, is an anti-ulcer drug. Shown are only compounds for therapeutic use. Diagnostics are not shown. If compounds serve both a therapeutic and a diagnostic purpose they are referred to as theragnostics.

treat rheumatoid arthritis. Bismuth(III) subcitrate (Figure 5.2) is used to treat gastric ulcers. Another type of metallodrug is tetrathiomolybdate. Its copper complexing capacity has been exploited in cancer therapy, because copper is necessary for angiogenesis (formation of new blood vessels). Copper sequestration limits blood supply and thus starves the tumour. Vanadium compounds are antidiabetic agents. There is significant interest in metallodrugs, in particular in extending the platinum pharmacology to ruthenium compounds, in finding new antibiotic agents, and in the use as radiopharmaceuticals.

Chelation. In metal overload conditions, it is not a matter of using a metallodrug but a chelating agent to remove the excess of metal. Chelation therapy is used to remove an excess of essential or non-essential metal ions. British anti-lewisite (BAL), dimercaptopropanol, DMSA, dimercaptosuccinic acid, and DMPS, dimercaptopropanesulfoxide are used as chelating agents for removing toxic metal ions. Iron overload is a serious medical condition that requires treatment with drugs such as deferoxamine, deferasirox or deferiprone.

5.2.2 Diagnosis

Unique physical properties of metal complexes make them suitable for medical imaging. Gadolinium(III) complexes are used in magnetic resonance imaging (MRI), and ^{99m}Tc compounds are used as radiopharmaceuticals in γ-ray imaging.

The use of metallodrugs and chelating agents makes it necessary to understand metallobiochemistry, the lifetime of these compounds in the body, their metabolism, their interaction with essential elements, and their modes of action.

5.3 The structures and functions of the non-essential metals

5.3.1 s-block alkali and alkaline earth metals

Li: The concentration of lithium in humans is 2–200 ng/g (wet tissue) with an uptake from diets of 60–70 μg/day. The toxicity from dietary sources is unknown. Giving 250–500 mg/d as Li_2CO_3 with blood plasma levels of 7–10 μg/ml is an effective treatment for manic depression, but there is very poor safety margin and a potential for kidney damage. The mechanism of lithium's action is not known. It affects many enzymes *in vitro.*

Be: Beryllium forms an insoluble hydroxide, $Be(OH)_2$, that prevents absorption in the intestinal tract. However, toxicity upon inhalation at the work place can lead to chronic beryllium disease, an immune-mediated lung disorder. Be is the antigen that elicits T-cell proliferation and sensitized T-cells and forms lung granulomas (berylliosis). The ensuing chronic lung inflammation leads to potentially fatal pulmonary fibrosis. Workers in the manufacturing of fluorescent lamps, the production of which has been discontinued, had been affected primarily. There is no effective method of decorporation.

Rb: Rubidium is quite abundant in human tissues. Its functions are not known.

Sr, Ba: Strontium and barium are also present at relatively high concentrations. Their functions are also not known.

5.3.2 d-block transition metals

Ti: The relatively high concentrations of titanium in the human body are noteworthy (Table 3.2). Research recognizing the presence and role of titanium in the human body is about 75 years old (Buettner and Valentine, 2012). Humans receive about 200 μg/day in the diet. The total amount in the human body is 10–20 mg with a concentration of about 2 μM in human blood. Ti is abundant in soil and, though less so, in sea water. The chemical species is Ti(IV), which can be solubilized by some biological ligands such as ascorbate. Experiments to induce deficiency were inconclusive.

Cd: Cadmium is a congener of zinc. It is classified as a human carcinogen. The sources of cadmium exposure are food and tobacco smoke. The toxicity of cadmium continues to be a public health concern. Exposure to cadmium damages the kidneys, the lungs, and effects on bone metabolism and the cardiovascular system have been documented. During at least half a century of research into the biochemistry underlying its toxicity many targets and pathways for the action of Cd have been identified. Cadmium(II) ions generate oxidative stress. The affinity of Cd for coordination environments with sulfur ligands is about three orders of magnitude higher than that of Zn. The replacement of zinc with cadmium in sites with sulfur ligands, however, does not satisfactorily explain its toxicity as structure and function can be preserved. However, if redox-inert cadmium replaces redox-active metal ions such as iron or copper, functions will be disrupted. Many cellular uptake mechanisms for cadmium have been identified. Interest in cadmium biochemistry began at a time when investigators were looking for the possible functions of additional metal ions and when metallothionein was isolated from horse kidney and found to contain significant amounts of cadmium in addition to other metal ions. Cadmium was then found to accumulate in metallothionein as a function of exposure and age. The interest further heightened when it was discovered that cadmium induces the synthesis of metallothionein. Based on these investigations, it was suggested that metallothionein detoxifies cadmium. A role in detoxification, however, is difficult to reconcile with the facts that (i) cadmium interferes with the function of zinc in this molecule, (ii) the induction actually occurs by an indirect mechanism in which cadmium displaces zinc in metallothionein and the released zinc binds to MTF-1 (metal regulatory element (MRE)-binding transcription factor-1) to induce the synthesis of metallothionein, and (iii) cadmium can be mobilized from metallothionein under conditions of oxidative stress. The numerous scientific investigations on cadmium need to be interpreted in terms of the actual concentrations in tissues under conditions of exposure.

Hg: Mercury is unique in being a liquid metal at ambient temperature and thus causing exposure through its vapours. The biochemistry relates mostly to Hg^{2+} (mercuric ion) and there are significant differences in uptake and toxicity depending on whether mercury is in inorganic or organic form such as monomethylmercury $HgCH_3^+$. Concerns continue about the health effects of environmental exposure to mercury and its compounds. Inorganic mercury is highly nephrotoxic. In addition, mercury affects the hematopoietic system and the central and peripheral nervous system. Mercuric ions form extremely tight bonds with the sulfur of thiols with dissociation constants as high as 10^{-20} M. The interactions with sulfur donors are thought to be major sites of action though the overall toxicity has not been linked to a single or a few molecular targets.

5.3.3 p-block metals

Al: Aluminium (AmE: aluminum) is the most prevalent element in the earth's crust. Aluminium ions are in the trivalent state. Very little Al^{3+} is available from insoluble $Al(OH)_3$ or $AlPO_4$, but some exists in a soluble form as $Al(OH)_4^-$. Importers, exporters and mechanism of homeostasis or detoxification are not known (Exley and Mold 2015). Yet, the metal is present at about the same total level (60 mg) as copper (70 mg) (Table 3.2). Aluminium binds to ferritin very strongly and unlike iron cannot be mobilized from it through oxidoreduction. There are measurable levels in blood (Table 3.2). There seems to be some accumulation in the brain, in particular an about two-fold increase with age. Aluminium exposure through various routes has been associated with dementia such as Alzheimer's disease. However, cause and effect are not clearly established.

Sn: Two μg Sn/g diet in rats did not show adverse symptoms, but when lowered to 17 ng deficiency symptoms were observed. The dietary tin intake in humans is around 1 mg/kg diet per day, e.g. from the tin leaching out from tin cans. The toxicity of inorganic and organic tin compounds is different. Organic tin compounds are neurotoxic.

Tl: As Tl_2SO_4 thallium has been widely used as a rodenticide. Its high toxicity causes a serious problem in accidental poisoning for humans and animals. It is thought that the interference with potassium metabolism is the major route of its toxicity. Although the Tl^+ and K^+ ions have similar sizes Tl binds much stronger than K and has a rather high affinity to sulfur ligands.

Pb: Perhaps better than any other element, lead illustrates the problematic relationship of some metals with human exposure. Lead poisoning has at least 2000 years of history. Romans experienced poisoning due to using lead salts ('sapa') as sweeteners in drinks. Lead as an anti-knocking agent added to gasoline was banned. The safety limits for lead are now being lowered even further because behavioral criteria as functional tests in children suggested neurological deficits at exposure levels below the ones reflected in recent recommendations for upper lead blood concentrations. One major source of exposure is lead-based paint in old buildings. Aside from interacting with the committing enzyme in heme biosynthesis, it is thought that lead interferes with Ca^{2+} metabolism. More than 90% of the lead ends up in bone.

Bi: Bismuth has been used for a long time as an antibiotic, and it still in use for treating stomach (gastric) ulcers. The trivalent metal ion (Bi^{3+}) is taken up by passive diffusion, complexed by glutathione, and transported by MRP (multidrug resistance protein) transporter into cellular vesicles, where it is deposited as a solid sulfide complex (Li and Sun 2012; Hong *et al.* 2015). Because of this sequestration mode, it is relatively nontoxic to humans but toxic for glutathione-poor organisms.

5.3.4 f-block metals

Lanthanides: These rare earth elements are now more widely used in a number of industrial and medical applications, and hence there is iatrogenic (caused by the action of a physician, here through application of an imaging agent) and occupational exposure, but very little is known about biological effects (Pagano *et al.* 2015). They occur in the trivalent state and are antagonists of calcium.

We are exposed to rare earth and other rare elements due to new technologies and products. For example, Pr, Nd, Sm, and Dy are used in high performance magnets, Y, La, Ce (La, Ce also for batteries), Eu and Tb for lighting, Ga, In and Te for solar cells, Er for optical fibers, Hf and Ta for electronic devices, and the *platinum group* Ru, Rh, Pd and their higher analogues Os, Ir, and Pt as catalysts in fuel cells (Ridgway 2015). For most of these elements, there are virtually no investigations on how they affect life.

Actinides: The first four metals are naturally occurring and the latter man-made. All are radioactive, but for some the concern for chemical toxicity is greater than the concern for radiotoxicity.

Summary

While we may not know all the essential elements, there is even less knowledge about the biological, pharmacological, and toxic functions of the non-essential elements. Nutrition, pharmacology, and toxicology are complementary aspects of a continuous spectrum of actions. The elements essential for life are mostly in the upper part of the PSE ($Z < 35$). The lower part of the PSE ($Z > 35$) contains most of the non-essential elements. Many of them are present in organisms. In addition, organisms become exposed to them. As with the essential elements dose/response curves of non-essential metal ions are idealized and may have different shapes if functions other than life and death are addressed, and they also are species-specific. Specific molecular targets have been identified in some cases but overall, molecular actions rarely explain the global effects. Every element/metal has a unique and individual biochemistry. A general focus on just a few elements or just the essential elements in biochemistry is not productive. Many of the apparently non-essential elements are present in organisms with functional outcomes and interact with the metabolism of essential elements. Hence, *we need to understand the biological functions of all elements in the PSE and their interactions with organisms*. Exploration of this area is a strength of the metallomics approach. Quantitative analytical and functional investigations need to be employed and combined to establish whether or not concentrations addressed in bioinorganic chemical investigations in isolated systems are relevant to organisms.

General references

R.B. Martin (1986). Bioinorganic Chemistry of Metal Ion Toxicity, Metal Ions in Biological Systems, H. Sigel, ed., Vol. 20 Concepts of Metal Ion Toxicity, Marcel Dekker, New York and Basel, pp. 21–65.

Trace elements in human nutrition and health. World Health Organization, Geneva 1996.

Specific references

A. Cvetkovic, A.L. Menon, M.P. Thorgersen, J.W. Scott, F.L. Poole 2nd, F.E. Jenney Jr., W.A. Lancaster, J.L. Praissman, S. Shanmukh, B.J. Vaccaro, S.A. Trauger, E. Kalisiak, J.V. Apon, G. Siuzdak, S.M. Yannone, J.A. Tainer, and M.W.W. Adams (2010). Microbial metalloproteomes are largely uncharacterized. *Nature* 466, 779–784.

A. Ridgway (2015). The materials bonanza. *New Scientist* 225, 35–41.

C. Exley and M.J. Mold (2014). The binding, transport and fate of aluminium in biological cells. *J. Trace Elem. Med. Biol.* 30, 90–95.

E.J. Calabrese (2013). Hormetic mechanisms. *Crit. Rev. Toxicol.* 43, 580–606.

F.H. Nielsen (2014). Should bioactive trace elements not recognized as essential, but with beneficial health effects, have intake recommendations. *J. Trace Elem. Med. Biol.* 28, 406–408.

G. Pagano, M. Guida, F. Tommasi, and R. Oral (2015). Health effects and toxicity mechanisms of rare earth elements — Knowledge gaps and research prospects. *Ecotox. Environ. Safe.* 115, 40–48.

H. Li and H. Sun (2012). Recent advances in bioinorganic chemistry of bismuth. *Curr. Op. Chem. Biol.* 16, 74–83.

K.D. Mjos and C. Orvig (2014). Metallodrugs in medicinal inorganic chemistry. *Chem. Rev.* 114, 4540–4563.

K.M. Buettner and A.M. Valentine (2012). Bioinorganic chemistry of titanium. *Chem. Rev.* 112, 1863–1881.

K.R. Drinker and E.S. Collier (1926). The significance of zinc in the living organism. *J. Industr. Hygiene* 8, 257–269.

N.P.E. Barry and P. Sadler (2013). Exploration of the medical periodic table: Towards new targets. *Chem. Commun.* 49, 5106–5131.

P. Chellan and P. Sadler (2015). The elements of life and medicines. *Phil. Trans. R. Soc.* A373, 20140182.

Y. Hong, Y.-T. Lai, G. C.-F. Chan, and H. Sun (2015). Glutathione and multidrug resistance protein transporter mediate a self-propelled disposal of bismuth in human cells. *Proc. Natl. Acad. Sci. USA* 112, 3211–3216.

Chapter 6

Regulation of metal ions

Walter Bradford Cannon (1871–1945) is considered as a father of systems biology. He introduced the term *homeostasis* (greek: ὅμοιος homœos, "similar" and στάσις stasis, "standing still") for the regulatory physiology observed by Claude Bernard (1813–1878) that living organisms maintain an *internal milieu* that protects and provides stability for cells and tissues. Importantly, the concept of homeostasis applies to essential metal ions. The realization how tightly they are regulated in the body (systemic homeostasis) and in the cell (cellular homeostasis) is one of the major recent advances in the field. It demonstrates the full integration of metal biochemistry into the general biochemistry of organisms and is crucial for understanding essential, pharmacological, and toxic functions of metal ions.

Aside from the contribution of metal ions to numerous biological functions, the extensive control of metal ions with proteins and the use of metabolic energy (ATP and GTP) emphasize why metal ions are central players in biochemistry. The biological control of metal ions determines metallomes and metalloproteomes, and thus the inorganic biochemistry of cells and organisms. Metalloproteins account for 30–40% of all proteins and in the category of metalloenzymes, the percentage is estimated to be as high as 50%. Metal ions must be exquisitely controlled to avoid a deficiency or an excess, to supply metal-requiring proteins with the correct metal at the right time and in the right location, and last but not least to make signalling with metal ions such as Ca^{2+} and Zn^{2+} possible. The control of metal allocation to proteins becomes evident when considering heterobinuclear metal sites in some metalloproteins. These proteins need two different metal ions, which must be made available at a 1:1 ratio despite the fact that the overall concentrations of the two metal ions may be quite different. Examples are cytosolic superoxide dismutase (SOD) requiring one copper ion and one zinc ion per subunit or calcineurin requiring one iron ion and one zinc ion. *Metallation* has to be accurate because a considerable amount of metabolic energy is spent for the synthesis of a protein. If the availability of the metal cofactor or metal prosthetic group were to limit the function of a protein, the energy would be wasted. Mismetallation, however, does occur if the systems for controlling metal ions are not working correctly. It can have serious consequences. Under conditions of metal excess, deficiency or perturbation of homeostatic mechanisms, the wrong metal can populate the metal site of a metalloprotein. For example, under anemia — including lead poisoning because it inhibits heme synthesis — zinc is incorporated into protoporphyrin IX. Another aspect is misfolding and aggregation of proteins, which can occur if the metal is required for the structure of a metalloprotein. An excess of metal ions may cause new interactions with proteins, thus affecting the folding and aggregation of many other proteins that are not metalloproteins. Mismetallation is different from possible switching of metal ions. It is thought that under certain conditions a different metal can be used in a metalloprotein for a biological purpose, perhaps adjusting to different metabolic needs (Foster *et al.* 2014). Biological systems evolved mechanisms to avoid scrambling of metals ions, to select metals from nutrients containing a large number of different metals, and to direct them exactly to where they perform specific functions. We will start describing the principles underlying this control and then describe some of the molecular systems and mechanisms governing selectivity and specificity. While presenting this key aspect of metal metabolism, in which the coordination environments of proteins and protein dynamics have a central role, the information in this chapter is more involved and open-ended in terms of unresolved issues than in the previous chapters.

6.1 Control of systemic homeostasis

Systemic homeostasis involves hormones and their signal transduction. Hormones coordinate intestinal uptake and release of metal ions, transfer to and distribution from the liver to different organs, and/or secretion

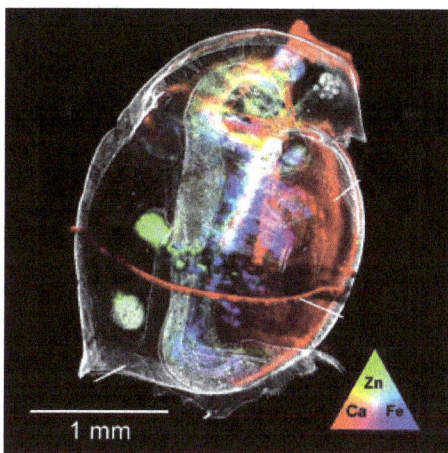

Figure 6.1 Metal distribution in a multicellular eukaryote. Distribution of zinc, iron, and calcium in a water flea (Daphnia magna) is uneven as determined by X-ray fluorescence microbeam techniques. From B. De Samber, R. Evens, K. De Schamphelaere, G. Silversmit, B. Masschaele, T. Schoonjans, B. Vekemans, C. R. Janssen, L. Van Hoorebeke, I. Szalóki, F. Vanhaecke, G. Falkenberg and L. Vincze (2013) *J. Anal. At. Spectrom.* **23,** 829; reproduced with permission.

by the kidneys. Hormones directing such metal traffic in the organism are known for only a few metals. For example, parathyroid hormone, vitamin D, and calcitonin regulate calcium; hepcidin regulates the export of iron from cells. Most metal ions have an uneven distribution in a tissue- and cell-specific way as shown for the water flea as an example (Figure 6.1). It is thought that all human cells require all the essential metal ions. The differences in distribution, therefore, may be quantitative rather than qualitative. While there has been enormous progress in understanding metal regulation at the cellular level, much more needs to be learned about organs and their interplay in regulating metal ions and how the uneven distribution in an organisms is brought about and controlled.

6.2 Selectivity and affinity: the Irving–Williams series

Biology has to cope with the inherent differences in stability of metal ion complexes. An overarching chemical principle to describe the stability of divalent transition metal ions including zinc is the Irving–Williams series (Irving and Williams 1948), according to which the stability increases in the following order:

$$Mn(II) < Fe(II) < Co(II) < Ni(II) < Cu(II) > Zn(II)$$

The reversed order at the end of the series indicates that Cu(II) complexes are generally stronger than Zn(II) complexes. The opposite sign is used in order to retain the order of the metals in the PSE. An expanded version of this series with the biologically significant divalent cations of magnesium and calcium is:

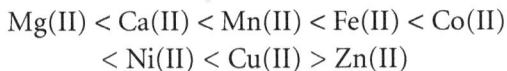

$$Mg(II) < Ca(II) < Mn(II) < Fe(II) < Co(II)$$
$$< Ni(II) < Cu(II) > Zn(II)$$

In biology, metal ions are present simultaneously and must be sorted so that a particular metal ion finds its place where it functions. The Irving–Williams series is an important principle for biology because the stability (affinity) dictates the free metal ion concentrations at equilibrium and therefore limits the range at which free metal ions are present. Vice versa, because affinities are an inherent property of the metal ions, it is necessary to keep the free metal ion concentrations of weaker binding ions high to saturate sites that require these ions. Thus, the free metal ion concentrations, i.e. metal ions not bound to proteins, are relatively high for Mg and Ca, but correspondingly low for the metal ions with much higher affinities such as Zn and Cu. There is a very important corollary for biology: *Each metal ion has a specific range of free metal ion concentrations, a "working range" where it can function without interference with the other metal ions* (indicated by arrows in Figure 6.2). These working ranges in which the available metal ion concentrations are controlled in cells are determined by several factors. Cobalt is kept in the vitamin B_{12} corrin cofactor, and nickel, as far as we know, is not used in metalloproteins in humans. When it is used in other organisms it is handled specifically by either a hydrocorphin cofactor or by a metallochaperone. The specific handling of cobalt and the apparent lack of physiological requirements for nickel in humans is important as it generates a gap and opens additional "working space" for the essential transition metal ions and zinc. It gives the two redox-inert metal ions calcium and zinc the possibility to fluctuate over a wider range

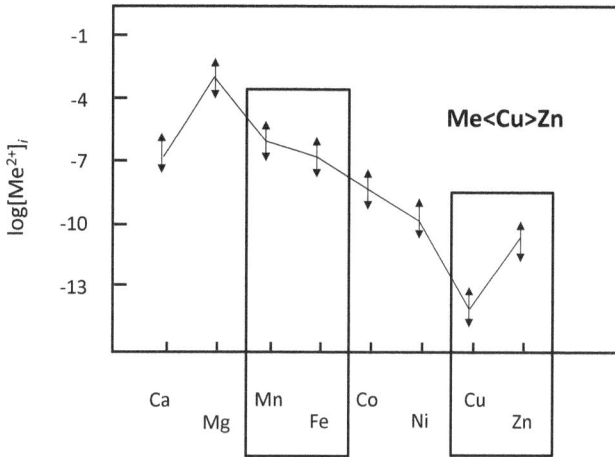

Figure 6.2   The Irving-Williams series. The series is extended to include the alkaline earth metals magnesium and calcium. The figure does not show affinities as a function of the divalent metal ions, which is the typical presentation for transition metal ions and zinc in most textbooks, but rather the range of free metal ion concentrations (arrows) expected from the affinities as a function of the metal ions. Despite the fact that metal ions become more thiophilic, i.e. have a preference for sulfur donors, in the series, the series seems to reflect the order in biology. There is one exception, though. Magnesium usually forms stronger complexes than calcium. However, in biology the order is reversed by providing additional stabilization in the coordination environments of calcium, which is needed for cellular signalling in eukaryota. The series illustrates the competition that biology has to control. The overall concentrations of the metal ions differ significantly in biology and this difference is an additional factor that biology has to deal with. Copper is an exception among the divalent metal ions as its major valence state in the cell is actually copper(I). From R.J.P. Williams (2002) *J. Inorg. Biochem.* 88, 241.

of concentrations and hence serve functions in cellular *signalling*. For these two metal ions, cellular homeostasis involves not only the maintenance of steady state concentrations but also the control of their fluctuations. Also, the gap sets apart the metal ions with relatively low affinity, Mn and Fe, from the ones with relatively high affinity, Cu and Zn. Hence, in humans it becomes mainly an issue of discriminating Mn from Fe on one hand and Cu from Zn on the other. Selectivity cannot be determined solely by affinities of metalloproteins for metals as is readily apparent when one considers the utilization of manganese. In humans, its overall concentrations are significantly lower than those of iron. Because it is the

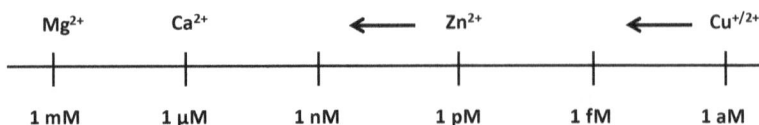

Figure 6.3 Free metal ion concentrations. A presentation that is different from the one in Figure 6.2 shows that metal ions can participate in biological regulation over 15 orders of magnitude in concentrations and thereby make an enormous contribution to cellular biology. The graph emphasizes the significance of the three redox-inert metal ions Mg, Ca, and Zn in biological control and illustrates how metal ions later in the Irving-Williams series, e.g. zinc and copper, affect physiological and pathophysiological events at vanishingly small concentrations. The effects of zinc are primarily due to binding (neglecting the fact that zinc affects the redox state indirectly), whereas the effects of copper are due to binding and its redox-activity, resulting in even higher cellular "toxicity" when its normal concentrations are perturbed.

least competitive transition metal ion in the series, how is it then possible that the more competitive metal ions do not gain access to its sites in proteins? This question will be addressed below.

For some metal ions the total and free metal ion concentrations are known with some degree of confidence. The concentrations of the free ions of the different metals cover over 15 orders of magnitude, and, when including potassium and sodium, even over 15 orders of magnitude (Figure 6.3). This scale is most remarkable as it allows for an unprecedented range of biological control, illustrating the enormous contribution metals ions make to the physiology of organisms.

The Irving–Williams series describes a trend. The actual affinities depend on coordination numbers and the types of ligands. The series is based on the same divalent metal ion concentrations and the same set of ligands. However, in biology, neither condition is met. The overall concentrations of each metal ion differ and for each metal there is not only one but several ligand sets to fulfil the different functions in metalloproteins. The coordination environments of proteins must be tuned in such a way that they bind a particular metal ion in the same range of affinities because the free metal ions need to be kept in the working range available. In addition to the variability in oxygen, nitrogen, and sulfur ligands, coordination numbers also vary. The consequence of such variations becomes significant for the affinities of catalytic metal ions. They have open coordination positions for binding substrate and therefore, on average, lower coordination numbers than structural sites. For example, catalytic zinc sites often have three protein ligands whereas structural sites have four. Thus, additional stabilizing factors must come into play in catalytic metal sites with lower coordination numbers to match the stability of sites with higher coordination numbers. One way to stabilize sites is hydrogen bonding from amino acids in the second coordination sphere to the donor ligands in the first coordination sphere. Moreover, this tuning of affinities for the purpose of biological control of metal ions must be achieved for each biological location where different constraints on the working ranges of the free metal ion concentrations pertain: outside cells, inside cells, and within cellular compartments. The coordination sites of metalloproteins are such that they do bind other metal ions in the test tube (Figure 2.10). However, in cells the availability of metals ions is tightly controlled and this control determines the binding of the correct metal ion to a metalloprotein. This interplay between the selectivity of protein sites for metal ions and the control of the availability of metal ions in certain ranges is the cornerstone of metal biology.

Other factors are also important in ascertaining selectivity. They include the control of the redox state of the transition metal ions iron and copper, which changes their coordination chemistry, and the handling of metal ions in prosthetic groups such as the heme group. Changing the redox state of a divalent transition metal ion takes the metal ion out of the equilibria dictated by the Irving-Williams series. Thus Fe(II) is brought into a thermodynamically much more stable state by oxidizing it to Fe(III). Copper is mostly copper(II) outside cells but primarily copper(I) in the cell. These factors further relax the constraints on metal ion concentrations and allow for a wider range of concentrations of some metal ions and thus additional functions. Extending the range is particularly important for metal ions such as calcium and zinc which participate in cellular signalling in multicellular eukaryota in addition to serving as cofactors of proteins. Extension of the working range for some metal ions is achieved by making available the range covered by cobalt and nickel in the Irving–Williams series through handling these metal ions by specific prosthetic groups and separating the handling of metal ions physically (in space) so that there is no chance of scrambling. Yet another way of controlling the working ranges is to use *metallochaperones* for highly competitive metal ions, e.g. copper and nickel. Metallochaperones do not solve the problem of selectivity of coordination environments because they use the same types of donor atoms from amino acid side chains to bind metal ions and their affinities are in the same range as their client proteins. However, metallochaperones provide additional metal ion control. They recognize their client proteins through protein–protein interactions and thereby increase the availability of metal ions locally. Overall they introduce additional buffering capacity and lower the concentrations of highly competitive free metal ions. The kinetic advantage is that they participate in associative metal transfer, thus metal transfer is independent of potentially slow dissociation rates from sites. Metallochaperones have mechanisms and coordination environments dedicated to the exclusive task of handling metal ions and avoiding the scrambling of metals. But it seems a "chicken-and-egg problem": If they do not have particular high selectivities for their metal ions, how is then scrambling avoided? The answer seems to be that they also interact with transport proteins at membranes where they are loaded with the correct metal ion, putting the burden of selectivity on the transporters and their control. Metallochaperones can be readily used for metals that have a limited number of client proteins. Zinc on the other hand, is also a rather competitive metal ion. About 3000 human proteins depend on it for their functions. Zinc is not thought to be distributed via metallochaperones as it would require a large number of metallochaperones or recognition sites for metallochaperones on each of the 3000 functionally and structurally different proteins. It is therefore thought that metallochaperones are used for the highly competitive copper to direct them to their sites in order to make it possible for zinc to be distributed to the many proteins without metallochaperones. In a protein such as cytosolic SOD with a 1:1 stoichiometry of copper and zinc in the active site, copper is inserted via a specific metallochaperone (CCS = copper chaperone for SOD) whereas zinc is believed to be supplied through the free zinc pool. Such a mechanism for two highly competitive metal ions avoids mismetallation because the metallochaperone inserts copper into the copper binding site.

Another class of metallochaperones are insertases which are employed for putting metal ions into the porphyrin and related prosthetic groups.

Ferrochelatase (protoheme ferrolyase) catalyzes the insertion of Fe^{2+} into protoporphyrin IX. Once the prosthetic group is metallated, ancillary proteins may insert it into the client protein.

While this section focused on affinities, i.e. thermodynamics, in the following section we will begin to discuss the contribution of kinetic factors in metal redistribution and in determining the time scale of biological events. The kinetics of metal association and dissociation also controls metal distribution. However, this area is much less explored. Protein environments modulate the kinetic behavior of metal ions, e.g. metalloenzymes. Metal ions may be kinetically trapped in a metalloprotein when the protein provides a coordination environment with extremely low dissociation rates for the metal ions. In metal sites that function in metal distribution, however, the protein provides a transient coordination environment to allow for fast dissociation. Transient metal-binding sites may have high affinities but conformational changes or other chemical reactions can make the metal ion available.

6.3 Metal buffering

By and large biological metal coordination follows the Irving–Williams series, which gives an indication in which range proteins buffer metal ions in a biological fluid, in a cell or in a cellular compartment. The concept of metal buffering is treated analogous to proton buffering and the definition of pH (1) with the free metal ion concentration expressed as pMe (2):

$$pH = -\log [H^+] \tag{1}$$

$$pMe = -\log [Me^{n+}] \tag{2}$$

While the Henderson–Hasselbalch equation (3) relates the pH value to the pK_a of the acid AH, the pMe value relates to the pK_d of the metal-ligand complex MeL (4):

$$pH = pK_a + \log ([A]/[AH]) \tag{3}$$

$$pMe = pK_d + \log([L]/[MeL]) \tag{4}$$

At equilibrium, there are metal-bound and metal-free forms of the ligands. Thus, the ratio between unbound ligand and metal-bound ligand determines the fraction of metal ion bound. Keeping a metal entirely bound, which is desirable for proteins that rely on a metal ion permanently for function, requires a certain amount of free metal ions to saturate the binding sites. As alluded to above, synthesizing proteins that are not saturated with a metal and hence not functional would be a waste of metabolic energy. The exception, of course, is proteins that function in metal buffering, sensing, and re-distribution. They do occur with different degrees of saturation. Biomolecules other than proteins also may serve as buffers to control metal ion concentrations. With few exceptions there is very little known which substances participate in buffering and whether or not specific substances serve as buffers for individual metal ions. If the same buffering species deals with different metal ions, an excess or a deficiency of one metal ion will affect the buffering of others.

Homeostatic control and the general principles of affinity expressed by the Irving–Williams series ascertain that the pMe for metal ions is rather constant in biological compartments. Under conditions of metal deficiency the metals are strongly buffered and less available, while under metal excess the metals are weakly buffered and available to bind to other targets. In addition to the pMe defining the free metal ion concentrations, the *buffering capacity* is an important parameter. It determines how resistant the pMe is to change, i.e. how strong the buffering is in a certain range around pMe. While buffering (pMe) depends on affinitites of ligands, buffering capacity depends on their concentrations. Changes of pMe are important if the demand for available metals increases in cellular processes and to allow for regulatory roles of metal ions. In principle, such changes could occur by changing the types of ligands or their concentrations. Switching to other types of ligands is employed in different biological compartments. When metal ions are released from subcellular compartments and from cells for the purpose of biological control, the buffering capacity affects the action of the metal ions. Under condition of high buffering capacity, it is difficult to effect changes of pMe. Buffering alone, therefore, is not sufficient as a mechanism to deal with the issue of making metals available and at the same time avoiding side effects of available free metal ions.

In a living system, the issue of buffering is more complicated. Living systems are in constant flux. Transport processes increase or decrease metal ion concentrations. Metal ions are exported from the cytosol, either out of the cell or into endosomal organelles (vesicles or vacuoles). They are also imported from outside the cell or from vesicles into the cytosol. This metal traffic helps maintaining the correct metal ion concentrations but it also changes metal concentrations of calcium and zinc required for biological control. This flux component of biological buffering has been referred to as *muffling* in the field of calcium biology (Thomas *et al.* 1991). This concept is more generally applicable. Muffling is limited by the capacities of transporters and storage of metal ions in vesicles. Thus, biological metal buffering is a combination of buffering due to the affinity of ligands, a thermodynamic component, and redistribution of metal ions, which adds a kinetic component. These concepts are the basis for the global responses to metal ions at different doses/concentrations (Chapters 4 and 5).

Whenever sulfhydryls are involved in metal binding there is a direct link between metal buffering and redox buffering as the oxidation of sulfhydryls reduces metal binding capacity of the sulfur donors whereas reduction of oxidized sulfur species to sulfhydryls increases metal binding capacity. Thus, redox changes affect metal ion availability when sulfhydryls are involved as donors. Redox reactions also control the valence state of redox-active metal ions and therefore can shift metal ions between weakly and strongly buffered species and vice versa, and this can be a major factor in controlling metal ion availability. For example, the free ferrous ion concentrations are relatively high but the free ferric ion concentrations are extremely low due to the much higher affinity of Fe(III) complexes compared to Fe(II) complexes. Moreover, redox-inert metal ions such as zinc also affect the redox state. Zinc binds to sulfhydryls and hence influences their reactivity towards oxidizing agents. In addition to physiological processes, this linkage between redox and metal homeostasis is important for many toxicological and pathophysiological conditions. Redox stress, either oxidative or reductive stress, thus affects metal ion availability and reactivity and can result in mismetallation, aggregation and misfolding of proteins.

Understanding and considering biological metal and redox buffering and their relationship is critical for planning experiments. Cells in culture do not have the selectivity that operates in an entire organism. The metal composition and buffering of culture media usually is significantly different from that encountered by cells in a biological tissue (or bacteria in their natural environments). Only if we know the ratio between total and free metal ion concentrations, we can buffer metal ion concentrations in media and in experiments with isolated molecules in a way that reflects the situation *in vivo*.

6.4 Control of cellular homeostasis

6.4.1 The proteins controlling homeostasis: coordination dynamics for metals on the move

At least five types of proteins participate in the control of homeostasis of essential metal ions: metalloregulators/sensors, membrane transporters, binding proteins, storage proteins, and metallochaperones (Figure 6.4),

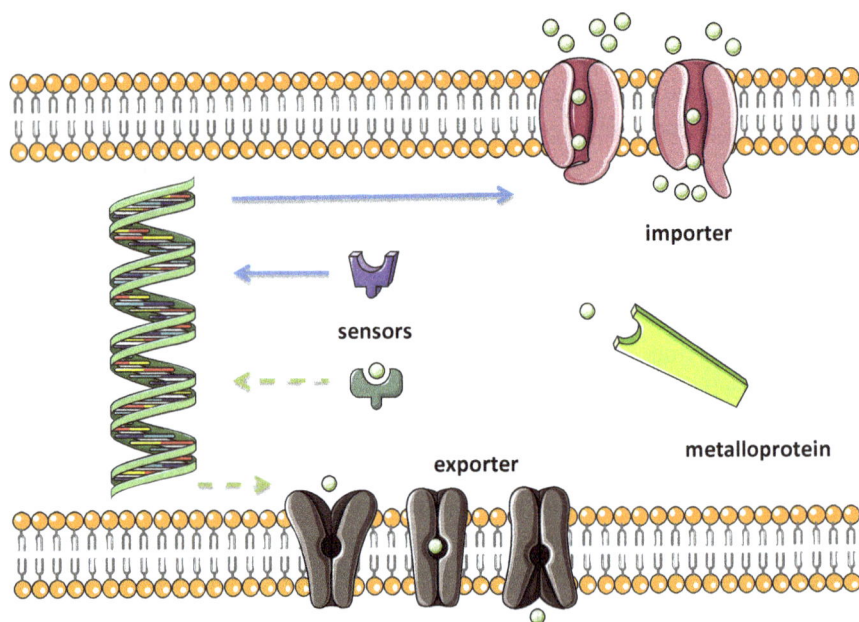

Figure 6.4   General principles of control and re-distribution of metal ions in cells. Key molecules are metal sensors, proteins that gauge whether free metal ion concentrations are too low or too high. They control transcriptional events — and translational events in eukaryotes — leading to changed membrane transporter activities for metal import or extrusion. Not shown are metallochaperones that transfer certain metal ions to proteins. The scheme (drawn with Servier Medical Art templates; www.servier.com) is highly simplified (compare to Figure 6.9).

most of which operate with mechanisms that are specific for individual metal ions. Feedback and feedforward control is required to adjust the correct total metal ion concentrations and the correct buffering to maintain the proper ratio between total and free metal ion concentrations. High up in the hierarchy are protein sensors that detect too high and too low metal ion concentrations and elicit specific transcriptional responses that involve the expression of proteins that take up, extrude and buffer metal ions. Together they determine the working ranges of metal ion concentrations from which functional metalloproteins obtain their metal ions.

In addition, regulation of translation, posttranslational mechanism of protein regulation, and regulation by miRNA integrate metal metabolism with other metabolic and signalling pathways in the cell. Metal ions must be made available or sequestered if the physiology of a cell changes, e.g. in proliferation, differentiation, and programmed cell death (apoptosis).

Unlike metalloenzymes which keep catalytic metals ions bound permanently during the lifetime of the protein, metal regulatory proteins bind metal ions transiently. When proteins regulate metals or metals regulate proteins, binding sites accept and deliver metal ions. Metal movement must occur without major changes in affinity as this would make sites susceptible to scrambling with other metal ions. A variety of mechanisms exist to endow sites with such coordination dynamics, which allows metal ions to move in proteins through channels or on the surface of proteins or to be transferred to another protein. Coordination dynamics of metal-binding sites with changes of structure and/or exchange of ligands are properties that are not directly evident from examining coordination environments. They are context-specific properties. In order to understand them, one needs to know the conditions for movements, cellular concentrations of metal ions and the pMe values for buffering. Such information is obtained by methods of chemical biology that determine conditions *in vivo*.

6.4.2 Selectivity filters: metal transporters

Coordination environments in metalloproteins are often far from absolute in terms of discriminating different metal ions. Many of them have limited selectivity when investigated *in vitro*. Even *in vivo*, for some metalloproteins, there is limited protection against the incorporation of a non-native metal ion. Thus, the lower affinity sites are prone to be attacked by higher affinity metal ions, in particular magnesium sites by zinc and iron-sulfur sites by copper. These statements about selectivity do not contradict the fact that proteins are highly evolved for their structure and function with the correct (native) metal ion. It merely indicates that control of metal ions is necessary and takes place "upstream" of the coordination environments of the proteins that are the ultimate recipients of the metal ion, namely at the level of the proteins that control the availability of metal ions. Membranes between compartments provide barriers for ionic species and specific transporters provide selectivity filters for import and export of metal ions. More specifically, the term selectivity filter is used for the pores that membrane proteins form and that determine the permeability of metal ions. But the filters alone do not solve the issue of metal allocation in cells either. Additional control is provided by transcriptional regulators for which different names are employed in the literature: metal sensors, metalloregulators, and metal receptors. We will use the first term to emphasize their primary functions as sensors. They control the expression of transporters and binding proteins. The sensors have been investigated extensively in bacteria but proteins with similar sensing characteristics are not known for humans and the question of how metal sensing works in humans is a matter of intense research efforts. Transporters and sensors have exquisite biological specificity in metal recognition coupled to functional outcomes. Coupling means that other metal ions may bind if available but they either are not delivered or do not elicit the biological response that only the native metal ion elicits.

Until a metal ion reaches its final destination, multiple selectivity filters determine its uptake and distribution (Figure 6.5). Selectivity in its extrusion provides another level of control. Transport proteins in the compartment between two membranes, e.g. Gram-negative bacteria or mitochondria with an outer and an inner membrane and a periplasmic or intermembrane space also determine the handling of metal ions. Compartments differ in the type and concentrations of proteins. In our bodies, from the lumen of the gut, where food chemistry and competition with bacteria determines availability, metal ions cross the intestinal apical membrane, are transported across the intestinal cell, which has its own cellular homeostatic mechanisms, and then cross the intestinal basolateral membrane to reach

ligands in: lumen   blood/   cytosol   organelle
               extracellular space

MeL_1 MeL_2 MeL_3 MeL_4 MeL_5

transporters of: two intestinal   cellular   subcellular
                   membranes   membrane   membrane

(A)

Ligands in: environment   periplasm   cytosol

MeL_1 MeL_2 MeL_3

transporters of: two membranes

(B)

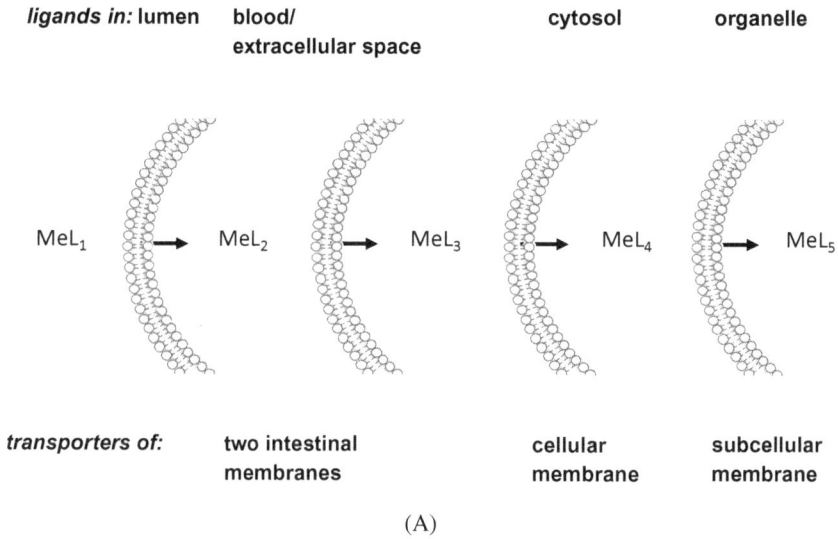

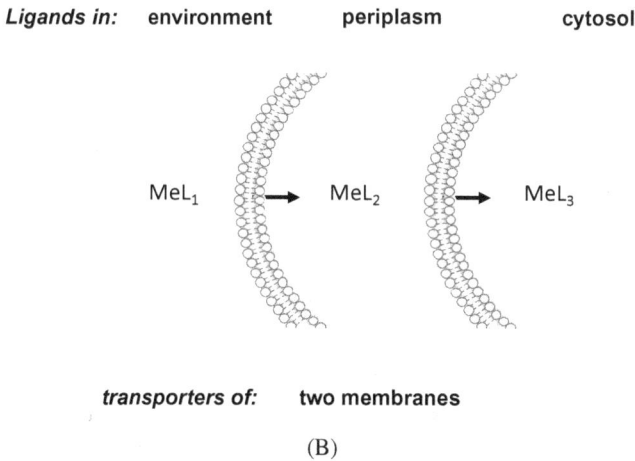

Figure 6.5   Membrane transporters as selectivity filters. The lipid bilayer of membranes generally is impermeable to ionic substances such as metal ions. Drawn with Servier Medical Art templates (www.servier.com). (A) In a eukaryotic organism, the metal ion has to pass through many membranes before it reaches its cellular or subcellular destination. The fluid spaces between the membranes provide at least five different ligands (L1-L5) and hence coordination environments that also have roles in the selectivity and sorting process. Additional barriers are provided to protect vital organs such as the brain. (B) Selectivity in a Gram-negative bacterial cell is determined by two membrane layers and three different ligands (L1-L3) outside the membranes.

the blood stream. The blood provides yet another set of ligands to bind metal ions. Some proteins in blood have specific roles in metal metabolism, such as transferrin for iron; others such as albumin bind multiple metals. The blood does not seem to have much selectivity regarding partitioning of metal ions to specific proteins. Aside from iron and transferrin, there is very little known about whether or not metal ions are loaded onto specific blood proteins at the intestinal cell and whether proteins other than transferrin are needed to supply cells with a particular metal ion. From the blood, metal ions cross the plasma membrane of cells. Finally, different subcellular membranes determine delivery to compartments within the cell. In this entire process of metal uptake and re-distribution, different ligand environments, changing valence states of the metal ion, and multiple transporters are involved. The changing ligand environment includes using methionine as a typical ligand for copper outside cells but cysteine as a ligand for copper inside cells. For zinc, oxygen and nitrogen coordination environments prevail outside the cell, while cysteine is a more common ligand inside cells. Coordination to the sulfhydryl group of cysteine is used inside the cell where there is a more reducing environment, while it is rarely used outside the cell where sulfhydryls are prone to oxidation. Sulfur coordination allows for sorting of thiophilic metal ions from others, i.e. Ca vs. Zn. Redox processes also play a role in selectivity. Fe^{2+} crosses membranes after reduction of Fe^{3+}; likewise Cu^+ is taken up after reduction of Cu^{2+}.

The selectivity of membrane transporters. Depending on their characteristics, protein transporters are classified as channels, energy(ATP)-requiring pumps, and carriers (diffusion facilitators). They can co-transport other species in the same (symporter) or opposite (antiporter) direction. The factors that lead to the exquisite selectivity of some transporters have been revealed for the alkali and alkaline earth metal ions. The similarity of Na^+, K^+, Mg^{2+}, and Ca^{2+} poses great challenges for generating selectivity. Chemical factors that make it possible for the proteins to discriminate among the four cations are fascinating, exemplary for the intricate bioinorganic chemistry of proteins, and critical for the fundamental biological functions of these metal ions. Potassium channels prefer potassium over sodium by a factor of 1000:1, while sodium channels prefer sodium over potassium by a factor of 500:1. The major factors of selectivity are the interplay between the protein matrix and the characteristics of the metal ion: binding energy and hydration energy of the cation, and the pore sizes, rigidity, solvent accessibility, and ligands that the protein provides to accommodate the metal ion (Dudev and Lim 2014). In some potassium channels, the metal-coordinating ligands are all carbonyls of the peptide bonds. The selectivity is in part determined by the preference of the potassium ion to adopt a coordination number of eight while the smaller sodium ion prefers six ligands. Sodium channels then work by selecting for the lower coordination number of sodium. They utilize aspartate and glutamate side chains in addition to carbonyls from the proteins' backbone. The discrimination between sodium and calcium also poses a major challenge as the two ions have nearly identical sizes. In calcium channels, this challenge is again met with 1000:1 selectivity of calcium over sodium using glutamate side chains. Yet a third challenge is the discrimination between calcium and magnesium. Magnesium with its higher charge density should form stronger complexes than calcium, but in biology, a higher affinity for calcium is provided, a particularly important issue as calcium is widely used in cellular signalling in eukaryota. A major factor in this discrimination is the coordination environment and coordination number. In EF-hand calcium binding sites in non-transporter proteins, the coordination is pentagonal bipyramidal (coordination number seven), which is better suited for the larger calcium ion than for the smaller magnesium ion, which prefers octahedral geometry (coordination number six). In other non-transporter calcium-binding proteins another factor comes into play. Calcium has a preference for the binding provided by the carboxyl groups of γ-carboxy glutamic acid (Gla), an amino acid derived from Glu with another carboxyl group introduced in a vitamin K-dependent reaction. This post-translational modification illustrates how the inventory of amino acids is enriched for the specific purpose of generating additional metal-interacting protein side chains.

The primary site of the selectivity filter is not the only site with a role in transport as the metal ion moves through the transporter protein and engages in interactions with other amino acids that serve as conduits for the metal. The transporters discussed here move the metal ions through the selectivity filter when the metals are devoid of their water ligands. However, the transporters must recognize chemical species other than the free ion inside and outside of the membrane and the removal of ligands and the reconstitution with ligands also contribute to transporter specificity.

Finally, transporters do not work in isolation. They interact with other biomolecules, and conformational changes and chemical modifications, such as phosphorylations, control the transport process. Membrane transporters have domains inside and outside the membrane. For the transition metal ions and zinc where interactions with ligands other than water are important, these domains have specific roles in ligand exchange.

Only recently, some insight into the membrane transport of d-group metals became available. Here, except for copper, which is transported as the monovalent ion, the divalent metal ions all have the same charge and very similar ionic radii, posing yet additional challenges. According to ligand preferences of d-group compared to s-group metal ions, additional amino acids such as nitrogen coordination from the His side chain are employed. One would expect that the Irving–Williams series applies. However, increasing affinity, which usually means lower dissociation rates, has to be counterbalanced by a sufficiently high rate of movement. The intrinsically higher affinity of transition metal ions and zinc compared to the alkali and alkaline earth metal ions requires environments that do not bind the metal too tightly in the transporter. While data are limited, the inherent stickiness of d-group metal ions appears to be overcome by using fewer ligands and more flexibility of ligands in coordination due to a lack of interactions with the secondary coordination sphere. One possibility is to employ lower affinities than dictated by the working ranges in the selectivity filter if the pore is sufficiently shielded to avoid interactions with competitive metal ions.

6.4.3 Metal sensors

Bacteria employ a very important component of metalloregulation: transcriptional metalloregulators/metal sensors. Many of these proteins are well characterized at the molecular level and, in contrast to membrane transporters for cations where knowledge of selectivity is primarily for alkali and alkaline earth metal ions, knowledge of selectivity for sensors exists for the entire series of essential d-group divalent cations. Similar to the membrane transporters with their selectivity filters, the sensors have exquisite selectivity and specificity. They are transcriptional activators (de-repressors), repressors or co-activators/co-repressors and work by three principles (triple "a" for mnemonics): affinity, access, and allostery (Waldron *et al.* 2009; Guerra and Giedroc, 2012), all of which are relative as they depend on the characteristics of the entire set of sensors for different metal ions present. Thus it is not just one factor, such as affinity that determines the response. Allostery (meaning binding at another site) elicited by metal ion binding results in conformational changes of the protein that determine DNA recognition. Access refers to kinetic factors in metal/sensor and ensuing sensor/DNA interactions. The sensors are the master molecules in the control of cellular metal ion concentrations. Significantly, the sensor characteristics describe quantitatively the working ranges of buffered, available metal ion concentrations in a cell. For example, the zinc-sensing system employs two different proteins that sense the thresholds for too low and too high zinc(II) ion concentrations. The set points match the affinities of the sensor proteins for zinc and trigger the expression of proteins for either zinc uptake or zinc efflux (Figure 6.6). There are

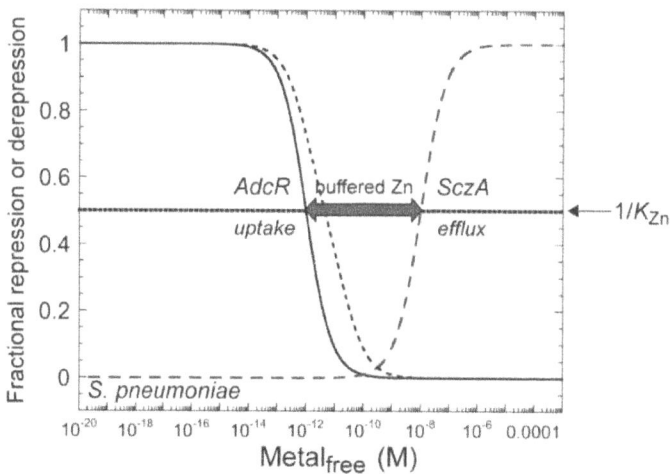

Figure 6.6 Bacterial zinc sensing. Two types of sensors for zinc in *Staphylococcus pneumoniae* are the arbiters of homeostatic control, one for sensing when the free metal ion concentrations are too low (AdcR) and the other when they are too high (SczA). The sensors a key to homeostatic mechanisms to control zinc in the correct ranges where the demand for it is satisfied and where adverse reactions are minimized. From A.J. Guerra and D.P. Giedroc (2012) *Arch. Biochem. Biophys.* **519**, 210; reproduced with permission.

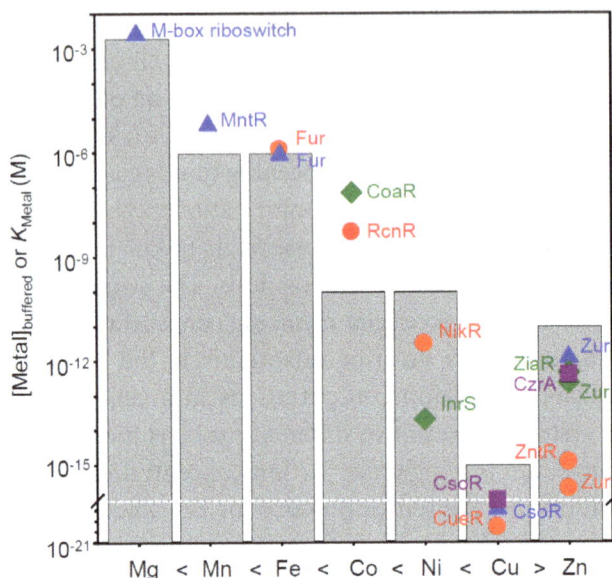

Figure 6.7 Bacterial metal sensing. Remarkably, the sensors for different metal ions respond exactly in the ranges predicted by the Irving-Williams series (see Figure 6.2), making the series a guiding general principle for biological control of metal ion concentrations. From A.W. Foster, D. Osman, and N.J. Robinson (2014) *J. Biol. Chem.* 289, 28095; reproduced with permission.

specific set points for the sensors of each metal ion and they follow the Irving–Williams series (Figure 6.7).

The metalloregulators detect the "free" metal ion concentrations, i.e. metal ions not bound to proteins. In iron biology the free metal ion concentration is referred to as the labile iron pool, which is thought to be in the range of micromolar concentrations (Hider and Kong 2013). The name "labile" makes some sense for Fe^{2+} which has relatively low affinity for proteins and hence dissociates significantly; however, it makes less sense for copper and zinc, which are tightly bound and do not dissociate significantly. Based simply on the magnitude of the equilibrium constant between protein-bound and non-protein bound metal ions, the term "free" metal ion is used albeit with the understanding that the metal ion has bound ligands. Thus, the meaning is different from the selectivity filters that transport metal ions devoid of any ligands. The ligand of iron in the labile iron pool is thought to be glutathione (Hider and Kong 2013). The ligands of free zinc and copper are not known.

Prokaryota have been investigated extensively with regard to metal regulation and some important principles have emerged. Cobalt and nickel are used in a limited number of processes and their pathways are linked to cofactor chemistries and control via metallochaperones. For the other transition metal ions and zinc, control is essentially a matter of competition between the four major ions, manganese, iron, copper, and zinc in two regions of affinity: Manganese and iron with relatively low affinity in the micromolar to nanomolar range and copper and zinc with relatively high affinity in the picomolar to femtomolar or even attomolar range. Thus, there are two issues, namely discriminating between high and low affinity binding metal ions, avoiding that high affinity metals populate low affinity sites (e.g. Mn, Fe vs. Cu, Zn), and discriminating between metal ions binding with similar affinities (e.g. Mn vs. Fe and Cu vs. Zn). One way of achieving the first objective is to keep the free (available) metal ion concentrations at very different levels as discussed above and another way is to separate the processes spatially. For example, with a free iron concentration of about 1×10^{-6} M ($1\ \mu$M), an iron protein with a $K_d(\text{Fe}) = 1 \times 10^{-7}$ M would be 90% saturated. Likewise, at a free zinc concentration of about 1×10^{-10} M a zinc protein with a $K_d(\text{Zn}) = 1 \times 10^{-11}$ M also would be 90% saturated but with insignificant population of the iron protein with zinc only if the affinities of the iron protein for zinc and the zinc protein for iron are not too high. To optimize the separation, either the difference of four orders of magnitude of the $K_d(\text{Fe})$ and $K_d(\text{Zn})$ must be complemented by some selectivity of the proteins, i.e. each of the proteins having less than the "expected" affinity for its native metal ions, or additional steps of selectivity must operate. With regard to spatial separation, copper is mostly periplasmic in bacteria and thus localization and folding of the protein are a way of discriminating between manganese and copper binding to the same type of protein (Tottey *et al.* 2008). To achieve the second objective, namely to discriminate between metal ions binding with similar affinities, additional factors come into play. The high affinity binders copper and zinc are handled differently: copper sites are populated with the aid of copper metallochaperones using an associative mechanism of metal transfer, whereas metallochaperones for zinc are not known, suggesting that zinc is handled primarily via dissociative mechanisms. Discrimination between low affinity binders has been addressed in bacteria that depend on iron and manganese to different degrees.

Bacillus subtilis needs both iron and manganese for growth and has about the same amounts of iron and manganese, the so-called metal quota, 0.5 mM (Helman 2014). This bacterium, therefore, is an excellent experimental model to address the challenges faced in discriminating between iron and manganese. In contrast, the growth of *Escherichia coli (E. coli)* does not depend on manganese but on iron, though *E. coli* uses manganese (Imlay 2014). On the other hand, Lactobacilli and *Borrelia burgdorferi* (a spirochete and tick-borne parasite) use little iron and rely mostly on manganese.

In *Bacillus subtilis,* three metalloregulators are responsible for the control of iron and manganese: Fur (iron uptake regulator) senses iron sufficiency, MntR senses manganese sufficiency, and PerR senses the Fe/Mn ratio. Fur and MntR bind to DNA in their apoform and control the expression of genes that encode the transporters for the import of the metal ions. Once the sensors bind the metal ion, they dissociate from DNA repressing gene expression. Structurally different sensors are used for the two metal ions. MntR belongs to the family of Dtx/Ide Fe^{2+} sensors. In contrast, PerR uses a different principle: the metal activates the protein and binding of the metallated form to DNA initiates transcription. The K_d (Fe^{2+}) is 0.2 μM whereas the K_d (Mn^{2+}) is 2.8 μM. Thus, one factor in selectivity is affinity. Some selectivity is embedded in coordination environments. Manganese(II) is mainly coordinated to oxygen, whereas iron(II) is slightly softer and thus prefers coordination to nitrogen (histidine) and sulfur (cysteine). This transition in the type of coordination environment was also noticed for the two pools of the free metal ions. This chemical property is universal and not restricted to bacteria. Manganese is buffered by oxygen-containing ligands whereas glutathione with its sulfhydryl ligand buffers iron (Hider and Kong 2013).

Eukaryota. Metal sensing in eukaryota seems to work by somewhat different principles because the typical metalloregulatory proteins of bacteria are not present. For zinc, at least one sensor has been well characterized: Metal regulatory element (MRE)-binding transcription factor-1 (MTF-1). For iron, sensing mechanisms at both the translational and transcriptional level have been characterized. Spatial separation, of course, can be employed much more extensively in eukaryotes due to their subcellular compartmentalization. Eukaryota have a nuclear membrane, mitochondria (and/or chloroplasts) and many cellular compartments (Figure 6.8). Hence, many

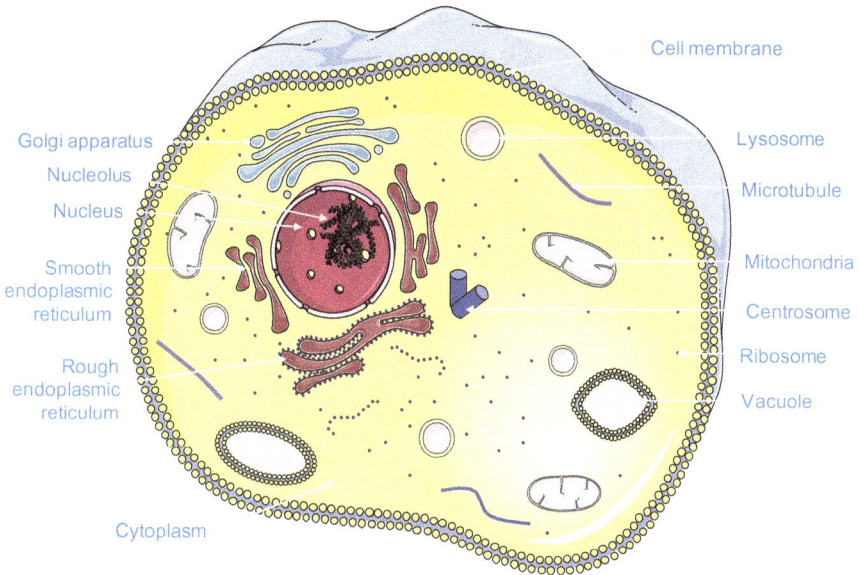

Figure 6.8 **Structure of a eukaryotic cell. A eukaryotic cell is highly compartmen-talized with membranes of organelles also serving as selectivity filters. Extensive intra-cellular traffic (transport) of metal ions occurs. Drawn with Servier Medical Art templates (www.servier.com).**

more aspects of metal regulation need to be considered. An important aspect of eukaryotic metal metabolism is cellular organelles/vesicles that are used for metal storage and release. Hence extensive intracellular re-distribution of metal ions occurs, adding considerable complexity to control but also providing new pathways for communication with metal ions between compartments and regulation of metabolism and signalling via metals ions such as calcium and zinc. Yeast (*Saccharomyces cerevisiae*) is widely investigated as a model eukaryote for metalloregulation but will not be treated here. Metalloregulation will be explained with iron and zinc as examples.

Iron

Iron homeostasis is tightly controlled, systemically by the hormone hepci-din (produced in the liver), which regulates iron export by ferroportin, and cellularly by transcriptional and posttranscriptional mechanisms.

Many regulatory aspects are known for human cellular iron metabolism, which is linked to oxygen metabolism. Iron needs to be channelled into three different pathways: (i) Fe–S cluster proteins, (ii) heme proteins, and (iii) non-heme iron proteins. The mitochondrial iron sulfur cluster (ISC) assembly and the cytosolic iron sulfur cluster assembly (CIA) machineries are dedicated to the first task. Ferrochelatase inserts iron into protoporphyrin IX to form heme. The insertion of iron into non-heme iron proteins may require iron metallochaperones, proteins that control the iron redox state, and proteins that stabilize a competent conformation of the recipient client proteins.

The posttranscriptional mechanisms were the first examples to show how iron and iron regulatory proteins (IRPs) control mRNA stability. A paradigm for translational regulation thus has emerged from investigation of iron metabolism. Iron-regulatory elements (IREs) are present in the 3′ or 5′ untranslated regions (UTR) of the mRNA of proteins involved in iron metabolism. Two types of IRPs bind to the hairpin structures of the IREs: iron-regulatory protein 1 (IRP1) and 2 (IRP2). These proteins serve as iron sensors. IRP1 is inactive when it binds a 4Fe-4S cluster and becomes a cytoplasmic aconitase (Aco1). IRP2 is regulated by degradation through an E3 ubiquitin ligase complex. A critical protein in this complex is the protein FBXL5, which is stabilized by iron and oxygen in a hemerythrin-like binuclear iron site (Kuehn 2015). Under iron-replete conditions IRP1/2 do not bind to the 5′ IRE hairpins and ferritin and ferroportin mRNAs are translated for iron storage and export from cells. Under iron-depleted conditions, IRP1/2 bind to 3′ IRE hairpins and transferrin receptor and DMT-1 mRNAs are translated for supplying cells with iron. Both IRPs act to increase or decrease the labile (free) iron pool (LIP) by controlling the mRNAs of transferrin or ferritin. There is additional regulation by heme sensing and transcriptional regulation. Posttranscriptional regulation involves proline hydroxylases, which are iron enzymes, regulating hypoxia-inducible factors (HIFs), which are transcription factors. The hydroxylation leads to ubiquitination and degradation of HIFs. This regulation is interconnected with translational regulation through an IRE in HIF2α (Simpson and McKie 2015).

A metabolic chart of iron metabolism illustrates this level of integration and interdependence (Figure 6.9).

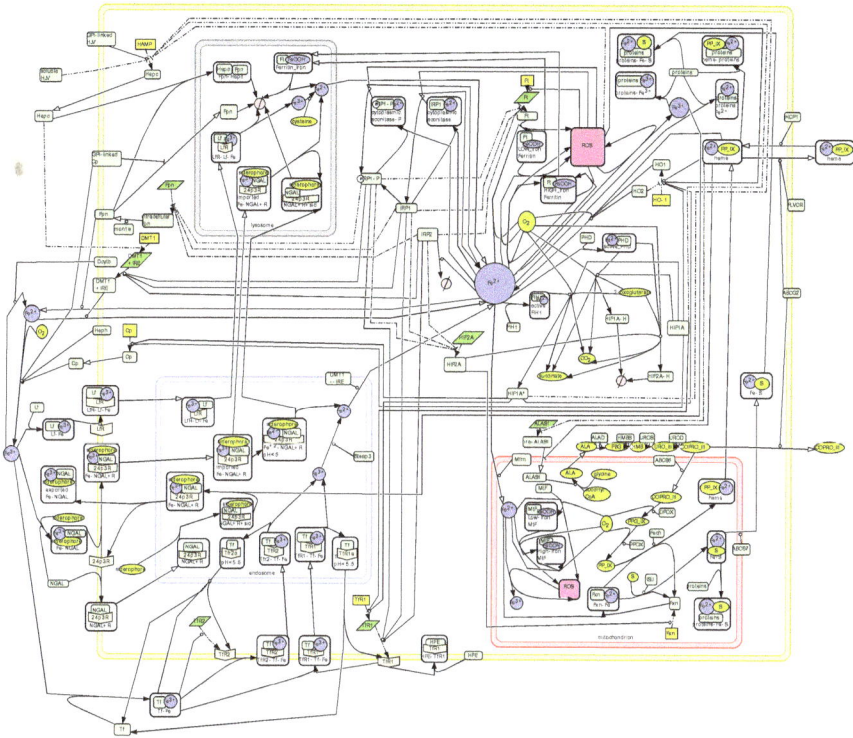

Figure 6.9   The iron metabolic network. The figure illustrates the integration of iron metabolism into cellular biology of a typical epithelial cell and the significance of a systems biology approach to describe the control of metal ions such as iron. The network includes 151 chemical species and 107 reaction and transport steps. Metals ions have direct and indirect actions. The complexity arises because metal ions affect proteins that are not metalloproteins and proteins that are not metalloproteins affect metal metabolism. The pleiotropic effects of metal ions are seen in transcriptomics investigations, where changing the metal ion concentrations affects a myriad of processes. From V. Hower, P. Mendes, F.M. Torti, R. Laubenbacher, S. Akman, V. Shulaev, and S.V. Torti (2009) *Mol. Biosyst.* **5**, 422; reproduced with permission.

Zinc

Many more proteins than for any other metal ion are involved in the control of zinc homeostasis. For zinc alone, several dozen of proteins coordinate directly the control of its cellular homeostasis in humans: 24 transporters of two families, Zip and ZnT (Kambe *et al.* 2015), about 12 metallothioneins, and at least one zinc-dependent transcription factor, MTF-1. MTF-1 senses high zinc(II) ion concentrations and induces the expression of proteins involved in zinc homeostasis (zinc transporters and metallothionein) and in other cellular functions. A corresponding sensor for low zinc concentrations is not known. The number of proteins involved in controlling zinc demonstrates how important the regulation of this essential metal ion is. Storage of zinc is different from that of iron for which a specific protein, namely ferritin, has evolved to store iron(III) in a solid iron oxide core. Zinc is stored in intracellular organelles as zinc ions, which can be made available from these organelles. An additional aspect of vesicular traffic is vesicular exocytosis. Zinc ions are secreted from cells to affect processes in the extracellular environment or in other cells. A number of cells secrete zinc. The best characterized cells in this regard are specific neurons that secrete zinc ions into the synaptic cleft, mammary gland epithelial cells that supply milk with sufficient zinc and pancreatic β-cells that load their granules with zinc for storage of crystalline insulin hexamers, which are held together by two zinc ions and one calcium ion and then secreted together with insulin.

6.4.4 Other transport and metal-binding proteins

Intracellular metallochaperones and chelatases/insertases are metal-binding proteins. Other proteins also participate in transport, delivery, and storage of metal ions. The nomenclature for these proteins is not clear-cut because either specific functions have not been identified or usage of specific terms has not yet developed.

Metal-binding proteins function inside and outside cells. Their selectivity for metal ions outside the cell can be less stringent as the mechanisms of cellular homeostasis apply only in as far as metal ions are taken up from the blood stream. Transferrin in the blood binds and transports metal ions other than its primary cargo, iron(III). Likewise proteins inside the cell involved in metal metabolism also bind several metal ions. The zinc-binding protein metallothionein binds copper and other metal ions. In fact, finding different metal ions in the isolated protein adduced to the general name metallothionein rather than calling it zinc thionein. The occupancy of these proteins with metals *in vivo* is presently being addressed. The fact that their metal-binding sites are not fully occupied and that they may pick up metal ions during isolation complicates the conclusions drawn about their metal binding selectivity. Similiar considerations apply to the major blood protein albumin.

Transferrin is primarily an extracellular transport protein. It binds Fe(III) with very high affinity in two binding sites. It binds to a receptor on the surface of cells and is taken up with the receptor by receptor-mediated endocytosis. The iron is then released in endosomes by a process that is enhanced by bicarbonate binding and acidification. Fe(III) is then reduced to Fe(II). This uptake of metal ions with their entire metal-binding protein employing a surface receptor is also a major part of the control of homeostasis.

Metallothionein can also be considered a transport protein for several metal ions, but primarily an intracellular one for zinc ions, although it also is secreted from cells and taken up by cells using a receptor/surface binding protein. Metallothionein transports zinc ions to the nucleus, apparently in an energy-requiring process, and to the intermembrane space of mitochondria. It also has specific mechanisms of delivering its metal cargo. Seven zinc ions are bound to the sulfur donors of 20 cysteine residues in two types of clusters that employ some of the sulfur donors as bridging ligands (Figure 6.10A). The sulfur (thiolate) donor allows for redox reactions of the otherwise redox-inert zinc ion. Coupling to cellular oxidants and oxidation of the sulfur releases zinc while reduction of the oxidized sulfur restores zinc binding capacity (Figure 6.10B).

Calmodulin has calcium-binding EF hands and is involved in calcium signalling. It binds four calcium ions, changes its conformation as a result of metal-binding, and in the calcium-loaded form binds to effector proteins such as certain kinases and activates them. There are other calcium-binding proteins that also serve as calcium-dependent effector molecules.

Ferritin is an iron storage protein (see above) with a protein shell and a nanocage of iron oxide in its core. Iron(II) binds and forms a binuclear iron site, where it is oxidized to iron(III) in an oxygen/hydrogen peroxide dependent process and then handed on for deposition as mineralized iron oxide. The mechanism of mobilization of iron from the core is not entirely clear but involves reduction of Fe(III) to Fe(II).

Zn₃S₉ cluster   Zn₄S₁₁ cluster

(A)

(B)

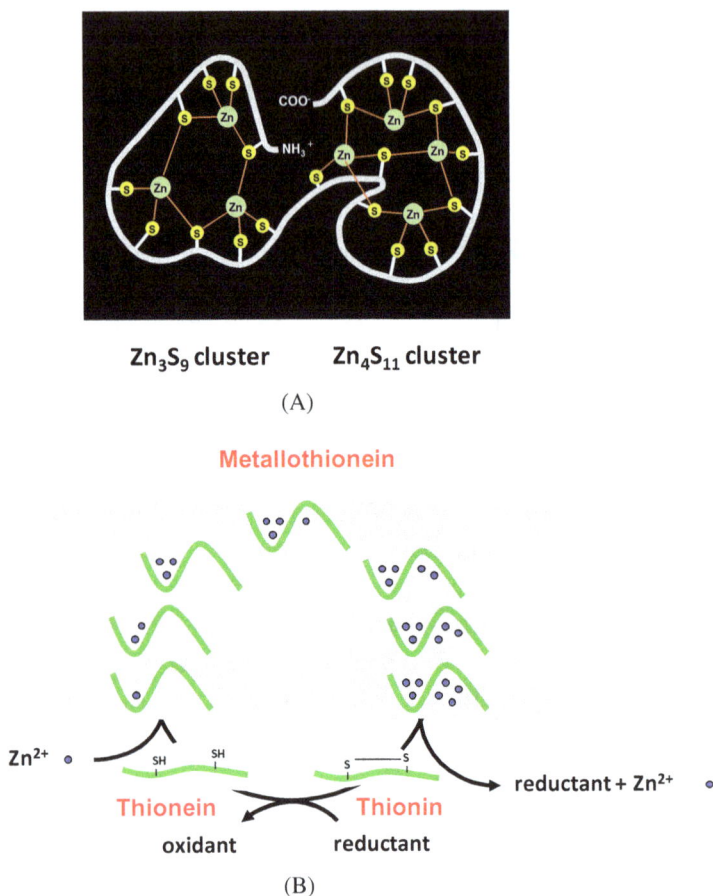

Figure 6.10   Metallothionein. (A) Mammalian metallothioneins (>10 different proteins) bind seven zinc ions using a total of 20 cysteines. This binding results in two clusters, one with three zinc ions bridged by three cysteine sulfur donors and the other with four zinc ions bridged by four cysteine sulfur donors. (B) Redox activity of metallothioneins. The sulfur donors confer redox activity on the clusters. Oxidation results in the release of the zinc(II) ions and the formation of the oxidized protein (thionin). Reduction of thionin forms thionein and restores zinc-binding capacity. From Y. Li and W. Maret (2008) *J. Anal. At. Spectrom.* 23, 1055; reproduced with permission.

In contrast, zinc and copper are stored in membrane vesicles in the cell. The principle is similar, though. In both cases the metals are removed from equilibria in the cytosol, in one case immobilized in a protein, and in the other case, sequestered in a vesicle.

6.5 Detoxification

Remaining questions to be addressed are what types of processes deal with non-essential metals and are there strategies for protecting organisms against toxic metal ions. Detoxification refers to processes that remove an excess of a metal ion by exporting it from cells or the organism or making it unavailable by sequestration. Export of a surplus of an essential metal ion from a cell is part of its homeostatic control and can be seen as a way of detoxification. Some non-essential metals are handled by the same systems that handle essential metals. Based on chemical similarity — referred to as molecular mimicry — non-essential or toxic metal ions may piggyback on the transporters for an essential metal. This process can be unidirectional and lead to the accumulation of metal ions. One can speculate that organisms do not have mechanisms to handle all metal ions. Protection may not exist against metal ions to which organisms usually are not or never have been exposed in their natural habitat. In this regard, the increasing exposure to toxic metal ions in our environment, including the exposure to "exotic" metals and nanomaterials provides new challenges. Detoxification is a relative term. An organism may be protected against a particular toxic metal ion but the metal ion may yet pose a threat to another organism in a food chain. Detoxification can be temporary or apparent only if the metal is not entirely removed from the organism and mobilized from its storage site. Organisms can be influenced by toxicants in more subtle ways without expressing overt acute toxicity.

Not all metal ions get readily into cells. Most of the metal transporters recognize divalent or monovalent cations or oxoanions and not cations in higher oxidation states. Once in the cell, toxic metals can be sequestered and made unavailable. If metals are deposited in some innocuous form in the cell, it is not necessarily without consequences for the organism. Pools of sequestered metal ions can be mobilized under stress conditions and then they become toxic. Chemical transformations of metal species can increase or decrease toxicity. Metal ions can be released from organometallic compounds or incorporated into organometallic compounds, thereby changing their toxicological profile. Many metal ions undergo redox reactions. In the cell, it is usually a reductive chemistry, which comes at the cost of reducing energy and may deplete species that are important for metabolism or detoxification of other compounds. Metal ions can induce oxidative stress by depleting cellular reductants, inhibiting enzymes involved in cellular reductions, or by direct generation of oxidative species. The resulting oxidative stress can then generate metal ions in higher oxidation states in the cell, at least transiently.

In this section, we will discuss briefly processes that have evolved to deal with some of the toxic metal ions. There are fundamental differences in the way bacteria and humans respond to toxic metal ions. While detoxification systems for some non-essential metals are well established in prokaryota, eukaryota seem to lack some of them. Instead, they rely more on barriers that deny toxic metal ion access into cells. Four systems for detoxification will be discussed:

(a) Membranes and barriers protecting against import
(b) Transporters for export
(c) Enzymatic and chemical transformations
(d) Metallothionein

Each system has a limited capacity and can be overwhelmed.

— Membranes and barriers. The various membranes and compartments that constitute the selectivity filters determine which metals gain access to cells. Additional barriers exist to protect vital organs to some extent. They guard mostly against ionic compounds but not uncharged, membrane-permeable compounds such as some organometallics.

— Transporters. Bacteria and plants can develop *resistance* to metal ions. They have inducible exporters (extrusion pumps) that remove the toxic metal from the cell. In bacteria, exporters are on the plasma membrane put in plants and other eukaryota, exporters are also located on intracellular membranes and remove metal ions from the cytosol into cellular vesicles where they may remain soluble or precipitate, e.g. lysosomes, aurosomes for gold. In some plants, storage of metals in cellular vacuoles can lead to hyperaccumulation of some metal ions. This process is employed in phytoremediation to remove metals from contaminated soils. Any species that depends on such plants for food runs a high risk of being poisoned.

— Transformation. Some enzymes use metal compounds as their substrates. It is yet another fascinating aspect of coordination dynamics as in this case it is not the metal in the active site performing the reaction but rather an active site interacting transiently with a metal ion during enzymatic catalysis. Bacterial organomercurial lyase (MerB) cleaves the Hg–C bond of organomercurials, generating Hg^{2+}. It acts in concert with mercury(II) reductase, the enzyme reducing the inorganic mercuric ion (Hg^{2+}) to elemental mercury. Both enzymes are part of the Mer operon, which also includes MerP, a periplasmic Hg^{2+} binding protein, and MerT, a membrane transporter for Hg^{2+}. The Mer operon is under control of the sensor for Hg^{2+}, MerR. This system brings both organomercurials and Hg^{2+} into the cell for the production of elemental mercury which is less toxic to the cell and then diffuses out of the cell. There are many more examples of enzymes using inorganic compounds as substrates, such as arsenic reductases or even enzymes reducing uranium compounds. Organomercurials accumulate in the fats of some fish and enter the food chain. Humans do not have an analogous mercury detoxification system. However, selenium compounds can provide some protection as selenide (Se^{2-}) forms highly insoluble HgSe. Forming insoluble materials is a strategy of decreasing toxicity. Likewise sulfides can precipitate bismuth compounds.

Essential and non-essential metal ions form glutathione complexes. These complexes are substrates for multidrug resistance associated proteins (MRPs), which export the complexes out of cells or into vesicles in an energy (ATP)-dependent process (Romero-Canelon and Sadler, 2015). A second step requiring metabolic energy is the induction of glutathione or metallothionein biosynthesis by some metal ions. Metals such as Bi, Pb, Ag are then deposited, presumably as insoluble sulfides, in intranuclear inclusion bodies. The process is important for resistance against metallodrugs such as those based on Pt and Au, but also for the selectivity of such drugs. For example, Bi is toxic to microorganisms but can be handled by human cells via the glutathione conjugation and deposition pathways.

— Metallothionein. Some metals induce the expression of metallothioneins, which then sequester the metal ions. The mechanism of cadmium induction of MTs involves cadmium binding to pre-existing zinc MT, displacing the zinc, which then binds to MTF-1 and activates the transcription of the protein (thionein), which then sequesters any surplus of cadmium. Though cited widely as a general mechanism of detoxification for cadmium, it is controversial whether cadmium is indeed detoxified in this way. Cadmium may interfere with the primary function of the MTs in zinc metabolism. Also, reactive species can mobilize cadmium from MT, which then affects other processes.

Summary

Metal ions are quite reactive and need to be tightly controlled in order for the essential metal ions to fulfil their functions in thousands of metalloproteins. Understanding the remarkable biological specificity of metalloregulation of essential metal ions and exclusion of non-essential metal ions requires knowledge of the coordination chemistry of metal ions and the architecture and dynamics of proteins. Control of metal ions occurs at the systemic and cellular levels of organisms. Selectivity filters, compartmentalization, metalloregulators, and coordination environments of the various proteins co-operate in providing the correct metal to make functional metalloproteins. Proteins involved in metalloregulation have specific molecular functions in transporting, storing, chaperoning, and sensing metal ions. The specificity of binding proteins, including metallochaperones, for metal ions is not absolute. Yet metalloregulation achieves remarkable specificity, in particular in the selectivity filters and sensors for metal ions. Nevertheless the coordination environments of metals in proteins are at the center of selectivity despite the fact that they are limited to using the ligands of only a few amino acid side chains. Metalloregulatory proteins have coordination dynamics to bind and deliver metals, requiring coupling between protein conformational changes and changes in coordination environments. The overarching principle underlying metalloregulation is that metals are buffered in typical working ranges of concentrations. Maintaining these concentrations is a prerequisite for correct metallation of a metalloprotein and for avoiding toxicity. Additional strategies include restricting reactions to cellular compartments and employing specific cofactors. Metallochaperones are used for competitive and redox-active metal ions and for assembling complex metal sites. They keep metal ions sequestered and thus away from unwanted reactions and they transfer metal ions to client proteins using molecular recognition in protein-protein interactions. Addressing bacterial metalloregulation has been instrumental in developing an understanding of fundamental aspects of the control of metal homeostasis. Key molecules are sensors that control gene expression of metal transporters. The affinities of the sensors cover the ranges of free metal ion concentrations typical for each metal ion in accordance with the Irving–Williams series. In eukaryota, there are additional steps in metal-loregulation involving extensive cellular re-distribution of metal ions among cellular organelles. Each subcellular compartment has its own metallobiochemistry. There are locally induced fluxes of metal ions from vesicles, in particular for the redox-inert metal ions calcium and zinc, both of which have roles in cellular signalling. Signalling occurs with locally fluctuating concentrations and requires additional steps in the control of homeostasis.

Non-essential and toxic metal ions challenge the homeostatic control of essential metal ions. Bacteria have sensors for toxic metals and efflux pumps to rid themselves of metal ions. Eukaryota do not have the types of resistance genes found in prokaryota. They restrict access of toxic metal ions to cells by barriers and selectivity filters and they sequester some toxic metal ions in proteins and in vesicles, precipitate them, or export them from cells.

Organisms control overall concentrations of each of a number of metal ions present simultaneously and the ratio between total and free metal ions. This situation differs from experiments performed ex vitro where single metal ions are investigated without buffering their concentrations. Metal coordination environments in proteins are not only optimized for function but also for affinities for metal ions in the range of allowable metal ion concentrations. Similar considerations apply to the kinetics of metal binding: Whether a metal site is permanent or transient depends on the biological requirements. The physicochemical properties of coordination environments become meaningful only in their biological context, highlighting uniquely biological aspects of coordination chemistry, which are not apparent from investigations of the isolated molecules. The functional attributes of structures derive from the requirements in the biological system.

References

A.J. Guerra and D.P. Giedroc (2012). Metal site occupancy and allosteric switching in bacterial metal sensor proteins. *Arch. Biochem. Biophys.* 519, 210–222.

A.W. Foster, D. Osman, and N.J. Robinson (2014). Metal preferences and metallation. *J. Biol. Chem.* 289, 28095–28103.

H. Irving and R.J.P. Williams (1948). Order of stability of metal complexes. *Nature* (London) 162, 746–747.

I. Romero-Canelon and P.J. Sadler (2015). Systems approach to metal-based pharmacology. *Proc. Natl. Acad. Sci. USA* 112, 4187–4188.

J.A. Imlay (2014). The mismetallation of enzymes during oxidative stress. *J. Biol. Chem.* 289, 28120–28128.

J.D. Helman (2014). Specificity of metal sensing: Iron and manganese homeostasis in *Bacillus subtilis. J. Biol. Chem.* 289, 28112–28120.

K.J. Waldron, J.C. Rutherford, D. Ford, and N.J. Robinson (2009). Metalloproteins and metal sensing. *Nature* 460, 823–839.

L.C. Kuehn (2015). Iron regulatory proteins and their role in controlling iron metabolism, *Metallomics* 7, 232–243.

R.C. Hider and X. Kong (2013). Iron speciation in the cytosol: An overview. *Dalton Trans.* 42, 3220–3229.

R.C. Thomas, J.A. Coles, and J.W. Deitmer (1991). Homeostatic muffling. *Nature* (London) 350, 564.

R.J. Simpson and A.T. McKie (2015). Iron and oxygen sensing: A tale of 2 interacting elements? *Metallomics* 7, 223–231.

S. Tottey, K.J. Waldron, S.J. Firbank, B. Reale, C. Bessant, K. Sato, T.R. Cheek, J. Gray, M.J. Banfield, C. Dennison, and N.J. Robinson (2008). Protein-folding location can regulate manganese-binding versus copper- or zinc-binding. *Nature* 455, 1138–1142.

T. Dudev and C. Lim (2014). Competition among metals ions for protein binding sites: Determinants of metal ion selectivity in proteins. *Chem. Rev.* 114, 538–556.

T. Kambe, T. Tsuji, A. Hashimoto, and N. Itsumura (2015). The physiological, biochemical, and molecular roles of zinc transporters in zinc homeostasis and metabolism. *Physiol. Rev.* 95, 749–784.

Epilogue and Summary

Under the cover of "Metallomics" the book discussed the position of the field of biometals with regard to chemistry and biology, the many opportunities that metallomics offers for other disciplines, and the importance of metals in biology for health and disease with a focus on humans and bacteria. Clearly, the subject matter can be expanded to include metalloids and non-metals in biological systems.

The importance of metal ions in biochemistry. Perhaps we have been tricked into believing that quantity is more important than quality and hence that some metal ions present at much lower amounts than bulk elements are qualitatively inferior to other elements. Zinc is a good example that such a perception is not correct. It is a micronutrient whose functions are as pleiotropic as those of some macronutrients. For humans, at least 10 metal ions are essential for human life. Their homeostatic control is critical for our health. Additional metal ions are essential for other forms of life. Metal deficiency, metal overload, and certain mutations of proteins that participate in the control of metal homeostasis cause disease. We need to know whether our status of essential metal ions is adequate and we need to understand metal metabolism when we employ metals or metallodrugs for therapeutic and diagnostic purposes. In addition, many non-essential metal ions affect our health.

The role of the periodic system of the elements in biochemistry. Metallomics addresses the roles of all metals in the periodic system of the elements, whether essential or non-essential. Major overarching issues are how organisms select elements and how the selected elements function in a particular biological system and interact with each other and non-essential elements. The simultaneous presence of many elements provides a view of biochemistry that is different from the traditional one which considers only a limited number of elements. Biochemistry needs to be seen in the context of all elements. The essential metals, Na, K, Mg, Ca, Mo, Mn, Fe, Co, Cu, and Zn are elements of life with a rich biochemistry that is firmly integrated into metabolism and cellular function and control. Zn, Fe, Ca, Mg, Na, and K affect numerous processes. The redox-inert divalent metal ions Mg, Ca, and Zn together participate in the control of biological processes over at least six orders of magnitude in concentrations. The role of Cu is slightly more specialized. The significance of Mn as an essential micronutrient is well established. Yet, it remains unknown how many proteins require it as a cofactor. Mo and Co have very special roles in only a couple of enzymes. The role of Cr is controversial. And then there are additional metals with roles presently established only in some organisms. Ni and V have no known molecular roles in humans but they have important roles in some organisms. The biological functions of other metal ions present in organisms and their implications for health and disease are less well understood.

Metals are more than just cofactors of proteins. Treating minerals and trace elements side-by-side with vitamins in biochemistry undermines their significance. Metal ions are associated mostly with protein function. At least 30–40% of all proteins contain a metal ion. With the exception of the interaction of Mg with RNA, the biological significance of metal ion interactions with other biomolecules is not clear. In addition to catalytic and structural roles in metalloproteins, metal ions participate in biological regulation and information transfer. Elaborate systems for homeostatic control have evolved, requiring proteins with a variety of functions. Metal ion binding increases protein stability and expands the types of shapes that proteins can adopt because metal ions organize structures of domains that apparently cannot be realized in their absence through the principles of protein secondary and tertiary structure. In addition to their roles in catalyzing chemically challenging reactions, in particular in the metabolism of gases, acquisition of metal binding sites is believed to have contributed to the evolution of life because the metal binding domains generated new interfaces for the interactions with other proteins and biomolecules and thus enhanced cellular organization and communication. The interaction between metal ions and proteins therefore is a major factor for normal and aberrant conformational landscapes of proteins and their aggregation. Proper metallation and mismetallation determine the balance between health and disease. Metal deficiency and overload compromise biological functions and provide multiple pathways to disease. We need to understand the metabolism and functions of essential metals and the mechanisms whereby metal toxicity deranges metabolism so that personal and public health can be optimized.

Coordination dynamics of biological metal sites. Many biological coordination environments of metal ions are dynamic, for example when metal ions interact with metalloproteins transiently. In enzymes, some metal ions change their coordination during catalysis. Proteins serve as conduits for metal ions during the extensive traffic of metal ions through transporters in the cytoplasmic membrane, in the cytosol, and during redistribution of metal ions among subcellular compartments. Metal association and dissociation and thus coordination dynamics is a central feature in the function of metalloregulatory proteins. Dynamic structures of biological metal complexes reflect the functions that proteins perform in the biological environment.

Metal ion control and buffering. Metals in cells and organisms are tightly controlled. Metal ion availability is completely different in molecular, cellular and animal experiments, posing challenges for performing biologically meaningful experiments *in vitro*. In a biological system, essential metal ions are available only in characteristic ranges of concentrations and they are all present simultaneously. Metal ion control determines which interactions with biomolecules and among metals are possible *in vivo*.

Metallomics — a systems approach to biometals. Metallomics has the potential of enriching the field of bioinorganic chemistry by addressing the additional roles that metal ions have when they function in biological systems. Metallomics was treated here as a general approach to systems and not as a set of methods. The functions of biometals in a system are *more than the sum of its parts: Functional significance of chemical structures is established in the context of the environment in which metal* *ions function. Questions such as "Why does a protein have a particular affinity for a metal ion and is the binding transient?" can be answered only from knowledge of the biology of the system.*

In contrast to traditional bioinorganic chemistry, which focuses on isolated molecules, a systems approach teaches us that the properties of molecules gain meaning in the context of the biological environment in which they function. There are plenty of examples of biomolecules having special, context-specific properties in biology. For example, the function of an isolated ATP molecule as a source of metabolic energy gains meaning only in the biological context when the energy of hydrolysis is harnessed in enzymatic reactions. The role of ester and anhydride bonds with phosphate in biological molecules, including the general role of ATP, became the foundation of biochemical energetics, but only sometime after the compounds were discovered. Likewise, metal coordination gains meaning when related to biological purpose and integrated with metabolism. Thus, the sulfur coordination environments of zinc in some metalloproteins allow for the redox-inert zinc to be coupled to redox reactions, a property evident only in the biological redox environment. Or another example: affinities of biological metal complexes must all be tuned to the specific control of metal homeostasis and their metabolic functions in the cell, again a property that gains meaning only through an understanding of the restrictions imposed by cells on metal concentrations and usage. The architecture of the coordination sites of metal ions in proteins has been optimized in terms of binding strength and kinetics to fulfil specific functions of an organism in the presence of other metal ions. For example, small or large differences in coordination chemistry of particular metalloproteins in different species are well established. However, explaining how these differences relate to function requires knowledge of the systems in terms of their demands on individual molecules.

The fascination with metals in biology derives from their remarkable structures and functions, which generated stimulating discussions in coordination chemistry. In biological systems, ligands from the macromolecule endow metal sites with characteristics unlike any of those previously known in chemistry because biological coordination environments evolved for specific purposes. When discovered, some structures were unprecedented in conventional chemistry. Once we know the structure we can attempt to imitate the coordination environments with synthetic chemistry. The expression and differentiation of biological function of metal sites is readily seen in the activation of metal ions for catalysis in metalloenzymes vs. the "deactivation" of a metal ion when it serves a structural function only. Bert L. Vallee (1919–2010) and Robert J.P. Williams (1926–2015) described metal ions in enzymes as being in an entatic state ($\varepsilon v \tau \varepsilon \acute{\iota} v \omega$ (*enteino*), "to stretch or strain tight", a term that was introduced in art history to describe the shape of columns in architecture) to call attention to this activation (Vallee and Williams 1968). The optimization of the structures of metal sites to biological function is probably what Berzelius meant by saying "to obey quite different laws" (see Introduction). The relationship between structure and function cannot be addressed in bioinorganic chemistry of isolated systems without knowledge of the functional context in the biological system. Conclusions about functions may be quite different depending on whether they are derived from an isolated compound or from the behavior of a compound in a system. A metal complex with high thermodynamic stability may be kinetically quite labile due to additional factors that affect reactivity, such as biochemically coupled reactions occurring in the biological environment. Or, as another example, an affinity of a metal to a macromolecule or a particular reactivity established *in vitro* may be outside the range of control taking place *in vivo* and hence not physiologically relevant. In trying to understand systems, metallomics has the potential to build a better understanding of the relationship between structure and function of the metals of life.

Reference

B.L. Vallee and R.J.P. Williams (1968). Metalloenzymes: The entatic nature of their active sites. *Proc. Natl. Acad. Sci. USA* 59, 498–505.

Glossary

Allostery: the process by which binding of a molecule (or ion) at one site regulates the activity of a macromolecule at another site.

Allotrope: Greek ἄλλος (allos) for "other" and τρόπος (tropos) for "form", different forms of an element with different structures, such as diamond and graphite for carbon.

Associative and dissociative mechanism: describes the transfer of metal ions from one coordination environment to another. In an associative mechanism the metal ion is never free as an intermediate with ligands from both coordination environments occurs; in a dissociative mechanism, the metal leaves its coordination environment completely before binding in another coordination environment.

Buffer: usually refers to chemical substances that bind protons making the system resistant to changes in pH. The concept can be extended to other substances such as metal ions. Metal buffers are chemical substances that bind metal ions and establish metal ion concentrations resistant to changes.

Cambialistic: describes the fact that some metalloenzymes can be active with different metal ions under different conditions.

Chelating agent: a chemical substance that can serve as a *ligand* and bind metal ions with several *donor* atoms.

Cofactor: a chemical substance that is required for the biological activity of a macromolecule (protein).

Donor: usually a non-metal that has electrons for engaging in bonding with a metal ion.

Endoplasmic reticulum: a compartment in eukaryotic cells — around the nucleus and sometimes extending to the plasma membrane — involved in many functions, primarily the handling of proteins for export into the extracellular space.

Endosome: a compartment in eukaryotic cells surrounded by a lipid bilayer membrane.

Entatic state: a state of tension (sterical strain) in the coordination environment.

Homeostasis: the control of internal conditions, e.g. in a cell or in blood. Metal homeostasis keeps metal concentrations in a certain range for physiology and thereby avoiding toxicity.

Hormesis: a concept describing a beneficial effect of otherwise toxic substances at low concentrations.

Kinetics: describes the *rates* of chemical reactions.

Ligand: a molecule that provides donor(s) for coordinating metal ions.

Metallation: providing a biomolecule such as a protein with the correct metal ion.

Metalloregulatory proteins: since functions of some of proteins and macromolecules in metal metabolism are still emerging, usage of terms is not consistent and inconsistencies can be confusing. The proteins include:

   Metal binding proteins: generic word for a metalloprotein with undefined function.

   Metallochaperones: proteins that bind a metal ion from another protein (donor) and deliver it to a specific protein that requires the metal ion for function (acceptor).

   Metallophores and ionophores: generally low molecular weight substances that sequester metal ions in an environment of poor metal ion availability and then bind to a receptor for uptake of the metal ions into cells. Siderophores carry Fe, chalcophores Cu, and zincophores Zn. *Ionophores* are chemical substances that are capable of carrying metal ions through membranes.

(**Metal**) **insertases:** proteins that insert metal ions into cofactors or prosthetic groups.

(**Metal**) **sensors:** also referred to as metal receptors in the earlier literature: proteins that gauge the cellular metal ion concentrations and elicit responses for adjusting the concentrations if they are either too low or too high.

(**Metal**) **storage proteins:** proteins specifically involved in sequestering metal ions to provide a pool of available metal ions for future demands.

(**Metal**) **transporters:** the term refers primarily to membrane-resident proteins involved in the uptake/influx and extrusion/efflux of metal ions (also used in combination with the terms channel, pump, carrier, symporter, antiporter, importer, exporter).

Metalloid: a chemical element that has the characteristics of both metals and non-metals.

Metal quota: the total concentrations of metal ions typical for a cell.

miRNA: micro RNA, small, non-coding RNA involved in silencing and posttranscriptional regulation of gene expression.

Prosthetic group: a substance that is tightly bound to a protein.

Redox potential: also reduction potential, characterizes a chemical substance in terms of how strongly it can accept or donate electrons and thereby reduce or oxidize.

Selectivity filters: the term refers to the arrangement of atoms in a membrane protein controlling whether or not a particular substance gains access to another compartment.

Thermodynamics: describes the *equilibrium* of chemical reactions.

Name and Subject Index

www.ingramcontent.com/pod-product-compliance
Lightning Source LLC
Chambersburg PA
CBHW061255220326
41599CB00028B/5660